Control Systems

Control Systems

W. Bolton

![Newnes logo]

Newnes

OXFORD AUCKLAND BOSTON JOHANNESBURG MELBOURNE NEW DELHI

Newnes is an imprint of Elsevier
Linacre House, Jordan Hill, Oxford OX2 8DP, UK
30 Corporate Drive, Suite 400, Burlington, MA 01803, USA

First edition 2002
Reprinted 2006

British Library Cataloguing in Publication Data
A catalogue record for this book is available from the British Library

Library of Congress Cataloging-in-Publication Data
A catalog record for this book is available from the Library of Congress

ISBN–13: 978-0-7506-5461-6
ISBN–10: 0-7506-5461-9

For information on all Newnes publications
visit our website at www.newnespress.com

06 07 08 09 10 10 9 8 7 6 5 4 3 2

Transferred to Digital Printing 2009

Contents

Preface

Aims   This book aims to provide an insight into the basic principles of control engineering and how these can be used to model the behaviour of such systems. The intention is to provide a book which considers in every-day language the principles involved with control systems without excessive emphasis on mathematics; familiarity with calculus notation and algebraic dexterity has, however, been assumed. As such, the book is ideal for higher level vocational courses and first studies of the topic at degree level.

The content has been carefully matched to cover the latest UK syllabuses, in particular the new specifications for BTEC Higher National in Engineering from Edexcel for the unit Control Systems and Automation.

Structure of the book   The book has been designed to give a clear exposition and guide readers through the principles of control engineering, reviewing background principles where necessary. Each chapter includes worked examples and problems. Answers are supplied to all the problems. The mathematics is introduced in the chapters as tools with a discussion of the principles of the mathematics relegated to two Appendices at the end of the book.

Content   The following is a brief overview of the content of the chapters in the book:

Chapter 1
This provides a non-mathematical overview of control systems, reviewing the concept of systems and their representation by block diagrams, how block diagrams can be used to represent open-loop and closed-loop control systems, examples of the measurement, signal processing and correction elements used in such systems and finally it is all brought together in a discussion of the structures of a wide range of examples of control systems.

Chapter 2
This introduces the development of models to represent systems and their representation by transfer functions.

Chapter 3
This follows on from Chapter 2 and uses the models to determine the responses of first-order and second-order systems.

Chapter 4

This is a continuation of Chapter 3 and looks at the parameters used to describe the performance of systems and the factors affecting stability.

Chapter 5

Following a review of the representation of sinusoidal signals by phasors, this chapter considers the response of systems to sinusoidal inputs, the frequency response function and the representation of the frequency response of systems by means of Bode diagrams. Stability and relative stability is discussed in relation to Bode diagrams.

Chapter 6

This chapter presents another method of considering the frequency response of systems, namely Nyquist diagrams.

Chapter 7

This chapter is a discussion of the controller and includes a consideration of on–off control by mechanical switching devices such as bimetallic strips and electronic switching by thyristors and transistor circuits, a descriptive and a mathematical consideration of PID control and tuning to obtain the optimum settings, concluding with a brief consideration of digital control systems.

Mathematics requirements

The following are the mathematics requirements for the various chapters:

Chapter 1

No mathematics is used in this chapter.

Chapter 2

This chapter uses calculus notation and shows how first and second-order differential equation models can be developed for systems. The differential equations have solutions given, the derivation of them being in an Appendix. The Laplace transform is introduced but used purely as a tool without any derivations from first principles. It is just used, following basic rules, as a means of transforming differential equations into algebraic equations.

Chapter 3

This uses a table of Laplace transforms and partial fractions, the rules for using them being given, to derive the outputs from systems. Essentially this chapter involves just algebraic manipulation.

Chapter 4

This requires algebraic dexterity in order to derive system parameters for first- and second-order systems.

Chapter 5

This uses, after introducing the principles, complex notation for phasors. Algebraic dexterity with equations involving j is then used to determine the frequency response of systems. The properties of the logarithm is assumed for the Bode plots.

Chapter 6

This introduces the polar graph and again requires algebraic manipulation of equations involving j.

Chapter 7

Very little mathematics is involved in this chapter, only in the section concerning controller mathematics is any mathematics involved and that is essentially just calculus notation.

W. Bolton

1 Control systems

1.1 Introduction The term *automation* is used to describe the automatic operation or control of a process. In modern manufacturing there is an ever increasing use of automation, e.g. automatically operating machinery, perhaps in a production line with robots, which can be used to produce components with virtually no human intervention. Also, in appliances around the home and in the office there is an ever increasing use of automation. Automation involves carrying out operations in the required sequence and controlling outputs to required values.

The following are some of the key historical points in the development of automation, the first three being concerned with developments in the organisation of manufacturing which permitted the development of automated production:

1   Modern manufacturing began in England in the 18th century when the use of water wheels and steam engines meant that it became more efficient to organise work to take place in factories, rather than it occurring in the home of a multitude of small workshops. The impetus was thus provided for the development of machinery.

2   The development of powered machinery in the early 1900s meant improved accuracy in the production of components so that instead of making each individual component to fit a particular product, components were fabricated in identical batches with an accuracy which ensured that they could fit any one of a batch of a product. Think of the problem of a nut and bolt if each nut has to be individually made so that it fitted the bolt and the advantages that are gained by the accuracy of manufacturing nuts and bolts being high enough for any of a batch of a nuts to fit a bolt.

3   The idea of production lines followed from this with Henry Ford, in 1909, developing them for the production of motor cars. In such a line, the production process is broken up into a sequence of set tasks with the potential for automating tasks and so developing an automated production line.

4   In the 1920s developments occurred in the theoretical principles of control systems and the use of feedback for exercising control. A particular task of concern was the development of control systems to steer ships and aircraft automatically.

5   In the 1940s, during the Second World War, developments occurred in the application of control systems to military tasks, e.g. radar tracking and gun control.

6   The development of the analysis and design of feedback amplifiers, e.g. the paper by Bode in 1945 on Network Analysis and Feedback Amplifier design, was instrumental in further developing control system theory.

7   Numerical control was developed in 1952 whereby tool positioning was achieved by a sequence of instructions provided by a program of punched paper tape, these directing the motion of the motors driving the axes of the machine tool. There was no feedback of positional data in these early control systems to indicate whether the tool was in the correct position, the system being open-loop control.

8   The invention of the transistor in 1948 in the United States led to the development of integrated circuits, and, in the 1970s, microprocessors and computers which enabled control systems to be developed which were cheap and able to be used to control a wide range of processes. As a consequence, automation has spread to common everyday processes such as the domestic washing machine and the automatic focusing, automatic exposure, camera.

This book is an introduction to the basic ideas involved in designing control systems with this chapter being an introduction to the basic idea of a control system and the elements used.

1.2 Systems

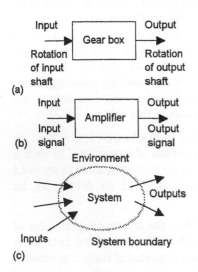

Figure 1.1   *Systems: (a) a gear box, (b) an amplifier, (c) the formal picture defining a system*

A car gear box can be thought of as a system with an input shaft and an output shaft (Figure 1.1(a)). We supply a rotation to the input shaft and the system then provides a rotation of the output shaft with the rotational speed of the output shaft being related in some way to the rotational speed of the input shaft. Likewise we can think of an amplifier as a system to which we can supply an input signal and from which we can obtain an output signal which is related in some way to the input signal (Figure 1.1(b)). Thus, we can think of a system as being like a closed box in which the workings of the system are enclosed and to which we can apply an input, or inputs, and obtain an output, or outputs, with the output being related to the input.

A *system* can be defined as *an arrangement of parts within some boundary which work together to provide some form of output from a specified input or inputs* (Figure 1.1(c)). The boundary divides the system from the environment and the system interacts with the environment by means of signals crossing the boundary from the environment to the system, i.e. inputs, and signals crossing the boundary from the system to the environment, i.e. outputs.

With an engineering system an engineer is more interested in the inputs and outputs of a system than the internal workings of the component elements of that system. By considering devices as systems we can concentrate on what they do rather than their internal workings.

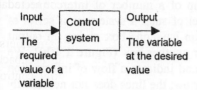

Figure 1.2 *Control system: an input of the required value of some variable and an output of the variable at the desired value*

Thus if we know the relationship between the output and the input of a system we can work out how it will behave whether it be a mechanical, pneumatic, hydraulic, electrical or electronic system. We can see the overall picture without becoming bogged down by internal detail. An operational amplifier is an example of this approach. We can design circuits involving operational amplifiers by making use of the known relationship between input and output without knowing what is going on inside it.

In this book we are concerned with control systems. *Control systems are systems that are used to maintain a desired result or value* (Fig. 1.2). For example, driving a car along a road involves the brain of the driver as a controller comparing the actual position of the car on the road with the desired position and making adjustments to correct any error between the desired and actual position. A room thermostat is another example of a controller, it controlling the heating system to give the required room temperature by switching the heater on or off to reduce the error between the actual temperature and the required temperature.

With a systems approach to control we express the physical system in terms of a model with the various physical components described as system blocks with inputs and outputs and the relationship between the inputs and outputs expressed by means of a mathematical equation.

1.2.1 Block diagrams

A useful way of representing a system is as a *block diagram*: *within the boundary described by the box outline is the system and inputs to the system are shown by arrows entering the box and outputs by arrows leaving the box.* Figure 1.3(a) illustrates this for an electric motor system; there is an input of electrical energy and an output of mechanical energy in the form of the rotation of the motor shaft. We can think of the system in the box operating on the input to produce the output.

While we can represent a control system as a single block with an input and an output, it is generally more useful to consider the system as a series of interconnected system elements with each system element being represented by a block having a particular function. Thus, in the case of the driver of a car steering the car along a road we can consider the overall control system to have the elements of:- the driver with an input of the actual position he/she sees of the car on the road and also his/her thoughts on where the car should be in relation to the road giving an output of the hands turning the steering wheel; the car steering unit with the input of the steering wheel position and the output of the front wheel positions and hence the positioning of the car on the road. Figure 1.3(b) shows how we might represent these elements.

In drawing formal block diagram models we use a number of conventions to represent the elements and connections:

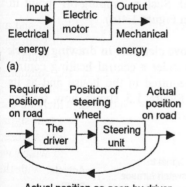

(a)

(b) Actual position as seen by driver

Figure 1.3 *Examples of block diagrams to represent systems: (a) an electric motor, (b) a car driving system involving a number of blocks*

1 *System element*

   A system element is shown as a box with an input shown as an inward directed arrow and an output as an outward directed arrow (Figure 1.4(a)).

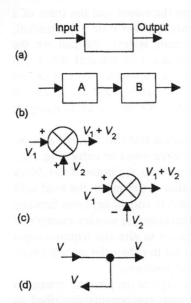

Figure 1.4 *Block diagram elements: (a) system element, (b) information paths, (c) summing junction, (d) take-off point*

2 Information flows

A control system will be made up of a number of interconnected systems and we can draw a model of such a system as a series of interconnected blocks. Thus we can have one box giving an output which then becomes the input for another box (Figure 1.4(b)). We draw a line to connect the boxes and indicate a flow of information in the direction indicated by the arrow; the lines does not necessarily represent a physical connection or the form of a physical connection.

3 Summing junction

We often have situations with control systems where two signals are perhaps added together or one subtracted from another and the result of such operations then fed on to some system element. This is represented by a circle with the inputs to quadrants of the circle given + or − signs to indicate whether we are summing two positive quantities or summing a positive quantity and a negative quantity and so subtracting signals (Figure 1.4(c)).

4 Take-off point

In the case of the car driving system shown in Fig. 1.3(b), the overall output is the actual position of the car on the road. But this signal is also tapped off to become an input to the car driver so that he or she can compare the actual position with the required position to adjust the steering wheel accordingly. As another illustration, in the case of a central heating system the overall output is the temperature of a room. But this temperature signal is also tapped off to become an input to the thermostat system where it is compared with the required temperature signal. Such a 'tapping-off' point in the system is represented as shown in Figure 1.4(d).

As an illustration of the use of the above elements in drawing a block diagram model for a control system, consider a central heating control system with its input the temperature required in the house and as its output the house at that temperature. Figure 1.5 shows how we can represent such a system with a block diagram.

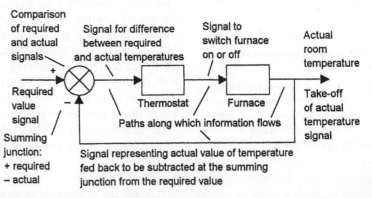

Fig. 1.5 *Block diagram for a central heating system employing negative feedback*

The required temperature is set on the thermostat and this element gives an output signal which is used to switch on or off the heating furnace and so produce an output affecting the variable which is the room temperature. The room temperature provides a signal which is fed back to the thermostat. This responds to the difference between the required temperature signal and the actual temperature signal.

1.3 Control systems models

There are two basic types of control systems:

1 *Open-loop*
In an open-loop control system the output from the system has no effect on the input signal to the plant or process. The output is determined solely by the initial setting. Open-loop systems have the advantage of being relatively simple and consequently cheap with generally good reliability. However, they are often inaccurate since there is no correction for errors in the output which might result from extraneous disturbances.

As an illustration of an open-loop system, consider the heating of a room to some required temperature using an electric fire which has a selection switch which allows a 1 kW or a 2 kW heating element to be selected. The decision might be made, as a result of experience, that to obtain the required temperature it is only necessary to switch on the 1 kW element. The room will heat up and reach a temperature which is determined by the fact the 1 kW element is switched on. The temperature of the room is thus controlled by an initial decision and no further adjustments are made. Figure 1.6 illustrates this. If there are changes in the conditions, perhaps someone opening a window, no adjustments are made to the heat output from the fire to compensate for the change. There is no information *fed back* to the fire to adjust it and maintain a constant temperature.

Open-loop control is often used with processes that require the sequencing of events by on–off signals, e.g. washing machines which require the water to be switched on and then, after a suitable time, switched off followed by the heater being switched on and then, after a suitable time, switched off.

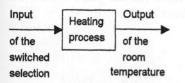

Figure 1.6 *Open-loop system with no feedback of output to modify the input if there are any extraneous disturbances*

2 *Closed-loop*
In a closed-loop control system a signal indicating the state of the output of the system is *fed back* to the input where it is compared with what was required and the difference used to modify the output of the system so that it maintains the output at the required value (Figure 1.7). The term *closed-loop* refers to the loop created by the feedback path. Closed-loop systems have the advantage of being relatively accurate in matching the actual to the required values. They are, however, more complex and so more costly with a greater chance of breakdown as a consequence of the greater number of components.

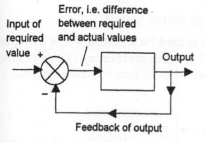

Figure 1.7 *Closed-loop system with feedback of output to modify the input and so adjust for any extraneous disturbances*

As an illustration, consider modifications of the open-loop heating system described above to give a closed-loop system. To obtain the required temperature, a person stands in the room with a thermometer and switches the 1 kW and 2 kW elements on or off, according to the difference between the actual room temperature and the required temperature in order to maintain the temperature of the room at the required temperature. In this situation there is *feedback*, information being fed back from the output to modify the input to the system. Thus if a window is opened and there is a sudden cold blast of air, the feedback signal changes because the room temperature changes and is fed back to modify the input to the system. The input to the heating process depends on the deviation of the actual temperature fed back from the output of the system from the required temperature initially set. Figure 1.8 illustrates this system with the comparison element represented by the summing symbol with a + opposite the set value input and a − opposite the feedback signal to give the sum as + set value − feedback value = error. This error signal is then used to control the process. Because the feedback signal is subtracted from the set value signal, the system is said to have *negative feedback*.

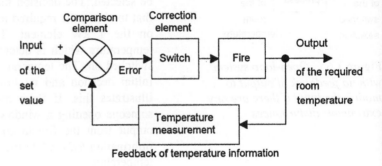

Figure 1.8 *Closed-loop system with feedback being used to modify the input to the controller and so enable the control system to adjust when there are extraneous disturbances*

1.3.1 Basic elements of an open-loop control system

The term *open-loop control system* is used for a system where an input to a system is chosen on the basis of previous experience as likely to give the output required. Figure 1.9 shows the basic form of such a system.

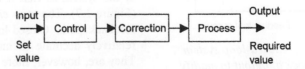

Figure 1.9 *Basic elements of an open-loop control system*

The system has three basic elements: control, correction and the process of which a variable is being controlled.

1   *Control element*
    This determines the action to be taken as a result of the input to the system.

2   *Correction element*
    This has an input from the controller and gives an output of some action designed to change the variable being controlled.

3   *Process*
    This is the process of which a variable is being controlled.

There is no changing of the control action to account for any disturbances which change the output variable.

1.3.2   Basic elements of a closed-loop system

Figure 1.10 shows the general form of a basic closed-loop system.

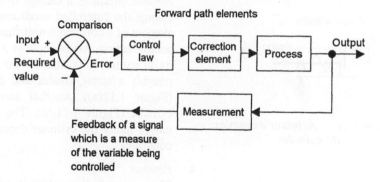

Figure 1.10 *Basic elements of a closed-loop control system*

The following are the functions of the constituent elements:

1   *Comparison element*
    This element compares the required value of the variable being controlled with the measured value of what is being achieved and produces an error signal:

    error = reference value signal – measured actual value signal

    Thus if the output is the required value then there is no error and so no signal is fed to initiate control. Only when there is a difference between the required value and the actual values of the variable will there be an error signal and so control action initiated.

2   *Control law implementation element*
    The control law element determines what action to take when an error signal is received. The control law used by the element may be

just to supply a signal which switches on or off when there is an error, as in a room thermostat, or perhaps a signal which is proportional to the size of the error. With a proportional control law implementation, if the error is small a small control signal is produced and if the error is large a large control signal is produced. Other control laws include *integral mode* where a control signal is produced that continues to increase as long as there is an error and *derivative mode* where the control signal is proportional to the rate at which the error is changing. These are discussed in Chapter 7.

The term *control unit* or *controller* is often used for the combination of the comparison element, i.e. the error detector, and the control law implementation element. An example of such an element is a differential amplifier which has two inputs, one the set value and one the feedback signal, and any difference between the two is amplified to give the error signal. When there is no difference there is no resulting error signal.

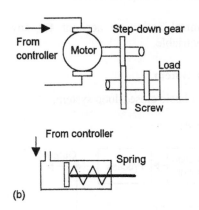

Figure 1.11   *Actuator examples: (a) motor, (b) cylinder*

3   *Correction element*
The correction element or, as it is often called, the *final control element*, produces a change in the process which aims to correct or change the controlled condition. The term *actuator* is used for the element of a correction unit that provides the power to carry out the control action. An example is a motor, with an input of a voltage to its armature coils and an output of a rotating shaft which, via possibly a screw, rotates and corrects the position of a workpiece (Figure 1.11(a)). Another example is a hydraulic or pneumatic cylinder (Figure 1.11(b)). The cylinder has a piston which can be moved along the cylinder depending on a pressure signal from the controller.

4   *Process*
The process is the system in which there is a variable that is being controlled, e.g. it might be a room in a house with its temperature being controlled.

5   *Measurement element*
The measurement element produces a signal related to the variable condition of the process that is being controlled. For example, it might be a temperature sensor with suitable signal processing.

The following are terms used to describe the various paths through the system taken by signals:

1   *Feedback path*
Feedback is a means whereby a signal related to the actual condition being achieved is fed back to modify the input signal to a process. The feedback is said to be *negative* when the signal which is fed back subtracts from the input value. It is negative feedback that is required to control a system. *Positive feedback* occurs when the signal fed back adds to the input signal.

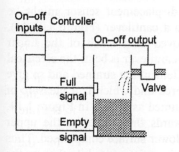

Figure 1.12 *Discrete-event control with the controller switching the valve open when empty signal received and closed when the full signal*

1.4 Measurement elements

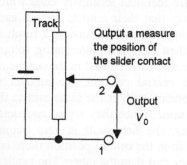

Figure 1.13 *Potentiometer as a sensor of position*

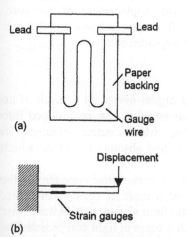

Figure 1.14 *(a) Strain gauge, (b) example of use on a cantilever to provide a displacement sensor*

2 *Forward path*

The term *forward path* is used for the path from the error signal to the output. In Figure 1.10 these forward path elements are the control law element, the correction element and the process element.

1.3.3 Discrete event control

This is often described as *sequential control* and describes control systems where control actions are determined in response to observed time-critical events. For example, the filling of a container with water might have a sensor at the bottom which registers when the container is empty and so gives an input to the controller to switch the water flow on and a sensor at the top which registers when the container is full and so gives an input to the controller to switch off the flow of water. This is a form of closed-loop system since the controller is receiving feedback from the two sensors regarding the state of the variable (Figure 1.12).

The following are examples of sensors that are commonly used with the measurement systems of control systems.

1.4.1 Potentiometer

A potentiometer consists of a resistance element with a sliding contact which can be moved over the length of the element and connected as shown in Figure 1.13. With a constant supply voltage V_s, the output voltage V_o between terminals 1 and 2 is a fraction of the input voltage, the fraction depending on the ratio of the resistance R_{12} between terminals 1 and 2 compared with the total resistance R of the entire length of the track across which the supply voltage is connected. Thus $V_o/V_s = R_{12}/R$. If the track has a constant resistance per unit length, the output is proportional to the displacement of the slider from position 1. A rotary potentiometer consists of a coil of wire wrapped round into a circular track or a circular film of conductive plastic over which a rotatable sliding contact can be rotated, hence an angular displacement can be converted into a potential difference. Linear tracks can be used for linear displacements.

1.4.2 Strain-gauged element

Figure 1.14(a) shows the basic form of an electrical resistance strain gauge. Strain gauges consist of a flat length of metal wire, metal foil strip, or a strip of semiconductor material which can be stuck onto surfaces like a postage stamp. When the wire, foil, strip or semiconductor is stretched, its resistance R changes. The fractional change in resistance $\Delta R/R$ is proportional to the strain ε, i.e.:

$$\frac{\Delta R}{R} = G\varepsilon$$

where G, the constant of proportionality, is termed the gauge factor. Metal strain gauges typically have gauge factors of the order of 2.0.

When such a strain gauge is stretched its resistance increases, when compressed its resistance decreases. A displacement sensor might be constructed by attaching strain gauges to a cantilever (Figure 1.14(b)), the free end of the cantilever being moved as a result of the linear displacement being monitored. When the cantilever is bent, the electrical resistance strain gauges mounted on the element are strained and so give a resistance change which can be monitored and which is a measure of the displacement. With strain gauges mounted as shown in Figure 1.14, when the cantilever is deflected downwards the gauge on the upper surface is stretched and the gauge on the lower surface compressed. Thus the gauge on the upper surface increases in resistance while that on the lower surface decreases.

1.4.3 Linear variable differential transformer

The linear variable differential transformer, generally referred to by the abbreviation LVDT, is a transformer with a primary coil and two secondary coils. Figure 1.15 shows the arrangement, there being three coils symmetrically spaced along an insulated tube. The central coil is the primary coil and the other two are identical secondary coils which are connected in series in such a way that their outputs oppose each other. A magnetic core is moved through the central tube as a result of the displacement being monitored. When there is an alternating voltage input to the primary coil, alternating e.m.f.s are induced in the secondary coils. With the magnetic core in a central position, the amount of magnetic material in each of the secondary coils is the same and so the e.m.f.s induced in each coil are the same. Since they are so connected that their outputs oppose each other, the net result is zero output. However, when the core is displaced from the central position there is a greater amount of magnetic core in one coil than the other. The result is that a greater e.m.f. is induced in one coil than the other and then there is a net output from the two coils. The bigger the displacement the more of the core there is in one coil than the other, thus the difference between the two e.m.f.s increases the greater the displacement of the core. Typically, LVDTs have operating ranges from about ±2 mm to ±400 mm and are very widely used for monitoring displacements.

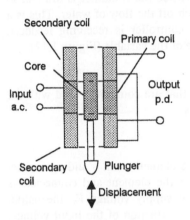

Figure 1.15   *LVDT: giving an output p.d. related to the position of the plunger*

1.4.4 Optical encoders

An encoder is a device that provides a digital output as a result of an angular or linear displacement. Position encoders can be grouped into two categories: incremental encoders, which detect changes in displacement from some datum position, and absolute encoders, which give the actual position.

Figure 1.16 shows the basic form of an *incremental encoder* for the measurement of angular displacement of a shaft. It consists of a disc which rotates along with the shaft. In the form shown, the rotatable disc has a number of windows through which a beam of light can pass and be detected by a suitable light sensor. When the shaft and disc rotates, a pulsed output is produced by the sensor with the number of pulses being proportional to the angle through which the disc rotates. The angular

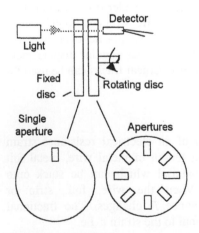

Figure 1.16   *Incremental encoder: angular displacement results in pulses being detected, the number of pulses being proportional to the angular displacement*

displacement of the disc, and hence the shaft rotating it, can thus be determined by the number of pulses produced in the angular displacement from some datum position. Typically the number of windows on the disc varies from 60 to over a thousand with multi-tracks having slightly offset slots in each track. With 60 slots occurring with 1 revolution then, since 1 revolution is a rotation of 360°, the minimum angular displacement, i.e. the resolution, that can be detected is 360/60 = 6°. The resolution typically varies from about 6° to 0.3° or better.

With the incremental encoder, the number of pulses counted gives the angular displacement, a displacement of, say, 50° giving the same number of pulses whatever angular position the shaft starts its rotation from. However, the absolute encoder gives an output in the form of a binary number of several digits, each such number representing a particular angular position. Figure 1.17 shows the basic form of an *absolute encoder* for the measurement of angular position. With the one shown in the figure, the rotating disc has four concentric circles of slots and four sensors to detect the light pulses. The slots are arranged in such a way that the sequential output from the sensors is a number in the binary code, each such number corresponding to a particular angular position. A number of forms of binary code are used. Typical encoders tend to have up to 10 or 12 tracks. The number of bits in the binary number will be equal to the number of tracks. Thus with 10 tracks there will be 10 bits and so the number of positions that can be detected is 2^{10}, i.e. 1024, a resolution of 360/1024 = 0.35°.

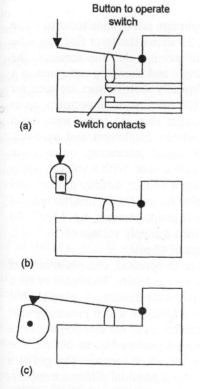

(a)
(b)
(c)

Button to operate switch

Switch contacts

Figure 1.18 *Limit switches: (a) Lever, (b) roller, (c) cam*

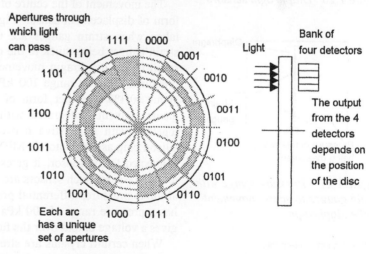

Apertures through which light can pass

1111 0000
1110   0001
1101   0010
1100   0011
1011   0100
1010   0101
1001   0110
1000 0111

Each arc has a unique set of apertures

Light

Bank of four detectors

The output from the 4 detectors depends on the position of the disc

Figure 1.17 *The rotating wheel of the absolute encoder: the binary word output indicates the angular position*

1.4.5 Switches

There are many situations where a sensor is required to detect the presence of some object. The sensor used in such situations can be a mechanical switch, giving an on–off output when the switch contacts are opened or closed by the presence of an object. Figure 1.18 illustrates the forms of a number of such switches. An example of switch application is

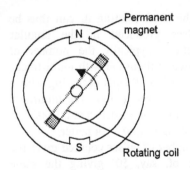

Figure 1.19   *The tachogenerator*

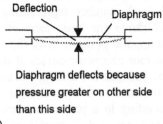

(a)

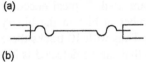

(b)

Figure 1.20   *Diaphragm sensors*

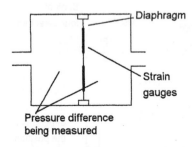

Figure 1.21   *Pressure gauge with strain gauges to sense movement of the diaphragm*

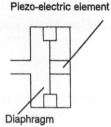

Figure 1.22   *Basic form of a piezo-electric sensor*

where a work piece closes the switch by pushing against it when it reaches the correct position on a work table, such a switch being referred to as a limit switch. The switch might then be used to switch on a machine tool to carry out some operation on the work piece.

1.4.6 Tachogenerator

The basic tachogenerator consists of a coil mounted in a magnetic field (Figure 1.19). When the coil rotates electromagnetic induction results in an alternating e.m.f. being induced in the coil. The faster the coil rotates the greater the size of the alternating e.m.f. Thus the size of the alternating e.m.f. is a measure of the angular speed.

1.4.7 Pressure sensors

The movement of the centre of a circular diaphragm as a result of a pressure difference between its two sides is the basis of a pressure gauge (Figure 1.20(a)). For the measurement of the absolute pressure, the opposite side of the diaphragm is a vacuum, for the measurement of pressure difference the pressures are connected to each side of the diaphragm, for the gauge pressure, i.e. the pressure relative to the atmospheric pressure, the opposite side of the diaphragm is open to the atmosphere. The amount of movement with a plane diaphragm is fairly limited; greater movement can, however, be produced with a diaphragm with corrugations (Figure 1.20(b)).

The movement of the centre of a diaphragm can be monitored by some form of displacement sensor. Figure 1.21 shows the form that might be taken when strain gauges are used to monitor the displacement, the strain gauges being stuck to the diaphragm and changing resistance as a result of the diaphragm movement. Typically such sensors are used for pressures over the range 100 kPa to 100 MPa, with an accuracy up to about ±0.1%. Another form of diaphragm pressure gauge uses strain gauge elements integrated within a silicon diaphragm and supplied, together with a resistive network for signal processing, on a single silicon chip as the Motorola MPX pressure sensor. With a voltage supply connected to the sensor, it gives an output voltage directly proportional to the pressure. Such sensors are available for use for the measurement of absolute pressure, differential pressure or gauge pressure, e.g. MPX2100 has a pressure range of 100 kPa and with a supply voltage of 16 V d.c. gives a voltage output over the full range of 40 mV.

When certain crystals are stretched or compressed, charges appear on their surfaces. This effect is called *piezo-electricity*. Examples of such crystals are quartz, tourmaline, and zirconate-titanate. A piezoelectric pressure gauge consists essentially of a diaphragm which presses against a piezoelectric crystal (Figure 1.22). Movement of the diaphragm causes the crystal to be compressed and so charges produced on its surface. The crystal can be considered to be a capacitor which becomes charged as a result of the diaphragm movement and so a potential difference appears across it. If the pressure keeps the diaphragm at a particular displacement, the resulting electrical charge is not maintained but leaks away.

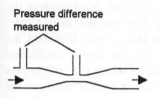

Figure 1.23 *Venturi tube: pressure difference is a measure of the flow rate*

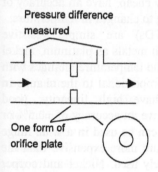

Figure 1.24 *Orifice plate*

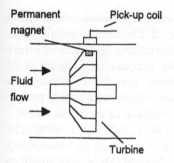

Figure 1.25 *Basic principle of the turbine flowmeter: number of pulses picked-up per second is a measure of flow rate*

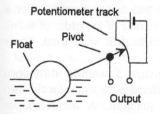

Figure 1.26 *Potentiometer float gauge: as the float rises the output p.d. decreases*

Thus the sensor is not suitable for static pressure measurements. Typically such a sensor can be used for pressures up to about 1000 MPa.

1.4.8 Fluid flow

The traditional methods used for the measurement of fluid flow involve devices based on Bernoulli's equation. When a restriction occurs in the path of a flowing fluid, a pressure drop is produced with the flow rate being proportional to the square root of the pressure drop. Hence, a measurement of the pressure difference can be used to give a measure of the rate of flow. There are many devices based on this principle. The *Venturi tube* is a tube which gradually tapers from the full pipe diameter to the constricted diameter. Figure 1.23 shows the typical form of such a tube. The pressure difference is measured between the flow prior to the constriction and at the constriction, a diaphragm pressure cell generally being used. The *orifice plate* (Figure 1.24) is simply a disc, with generally a central hole. The orifice plate is placed in the tube through which the fluid is flowing and the pressure difference measured between a point equal to the diameter of the tube upstream and a point equal to half the diameter downstream. Because of the way the fluid flows through the orifice plate, such measurements are equivalent to those taken with the Venturi tube.

The turbine flowmeter (Figure 1.25) consists of a multi-bladed rotor that is supported centrally in the pipe along which the flow occurs. The rotor rotates as a result of the fluid flow, the angular velocity being approximately proportional to the flow rate. The rate of revolution of the rotor can be determined by attaching a small permanent magnet to one of the blades and using a pick-up coil. An induced e.m.f. pulse is produced in the coil every time the magnet passes it. The pulses are counted and so the number of revolutions of the rotor can be determined. The meter is expensive, with an accuracy of typically about ±0.1%.

1.4.9 Liquid level

A commonly used method to measure the level of liquid in a vessel is a float whose position is directly related to the liquid level. Figure 1.26 shows a simple float system. The float is at one end of a pivoted rod with the other end connected to the slider of a potentiometer. Changes in level cause the float to move and hence move the slider over the potentiometer resistance track and so give a potential difference output related to the liquid level.

1.4.10 Temperature sensors

The expansion or contraction of solids, liquids or gases, the change in electrical resistance of conductors and semiconductors, thermoelectric e.m.f.s and the change in the current across the junction of semiconductor diodes and transistors are all examples of properties that change when the temperature changes and can be used as basis of temperature sensors.

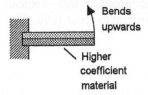

Figure 1.27 *Bimetallic strip; as the temperature increases it bends upwards with the higher coefficient material extending more than the lower coeffiicient material*

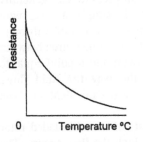

Figure1.28 *Variation of resistance with temperature for thermistors*

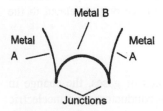

Figure 1.29 *Thermistors: (a) rod, (b) disc, (c) bead*

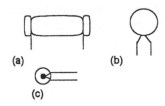

Figure 1.30 *Thermocouple: a temperature difference between the junctions gives a p.d. between them*

The *bimetallic strip device* consists of two different metal strips of the same length bonded together (Figure 1.27). Because the metals have different coefficients of expansion, when the temperature increases the composite strip bends into a curved strip, with the higher coefficient metal on the outside of the curve. The amount by which the strip curves depends on the two metals used, the length of the composite strip and the change in temperature. If one end of a bimetallic strip is fixed, the amount by which the free end moves is a measure of the temperature. This movement may be used to open or close electric circuits, as in the simple thermostat commonly used with domestic heating systems. Bimetallic strip devices are robust, relatively cheap, have an accuracy of the order of ±1% and are fairly slow reacting to changes in temperature.

Resistance temperature detectors (RTDs) are simple resistive elements in the form of coils of wire of such metals as platinum, nickel or copper alloys, the resistance varying as the temperature changes with the change in resistance being reasonably proportional to the change in temperature. Detectors using platinum have high linearity, good repeatability, high long term stability, can give an accuracy of ±0.5% or better, a range of about −200°C to +850°C, can be used in a wide range of environments without deterioration, but are more expensive than the other metals. They are, however, very widely used. Nickel and copper alloys are cheaper but have less stability, are more prone to interaction with the environment and cannot be used over such large temperature ranges.

Thermistors are semiconductor temperature sensors made from mixtures of metal oxides, such as those of chromium, cobalt, iron, manganese and nickel. The resistance of thermistors decreases in a very non-linear manner with an increase in temperature, Figure 1.28 illustrating this. The change in resistance per degree change in temperature is considerably larger than that which occurs with metals. For example, a thermistor might have a resistance of 29 kΩ at −20°C, 9.8 kΩ at 0°C, 3.75 kΩ at 20°C, 1.6 kΩ at 40°C, 0.75 kΩ at 60°C. The material is formed into various forms of element, such as beads, discs and rods (Figure 1.29). Thermistors are rugged and can be very small, so enabling temperatures to be monitored at virtually a point. Because of their small size they have small thermal capacity and so respond very rapidly to changes in temperature. The temperature range over which they can be used will depend on the thermistor concerned, ranges within about −100°C to +300°C being possible. They give very large changes in resistance per degree change in temperature and so are capable, over a small range, of being calibrated to give an accuracy of the order of 0.1°C or better. However, their characteristics tend to drift with time. Their main disadvantage is their non-linearity.

When two different metals are joined together, a potential difference occurs across the junction. The potential difference depends on the two metals used and the temperature of the junction. A *thermocouple* involves two such junctions, as illustrated in Figure 1.30. If both junctions are at the same temperature, the potential differences across the two junctions cancel each other out and there is no net e.m.f. If, however,

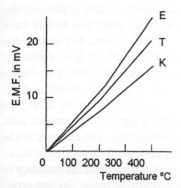

Figure 1.31 *Thermocouples: chromel-constantan (E), chromel-alumel (K), copper-constantan (T)*

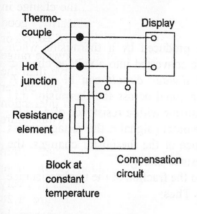

Figure 1.32 *Cold junction compensation to compensate for the cold junction not being at 0°C*

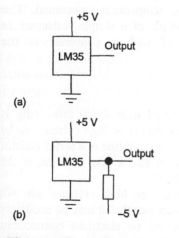

Figure 1.33 *LM35 connections*

there is a difference in temperature between the two junctions, there is an e.m.f. The value of this e.m.f. *E* depends on the two metals concerned and the temperatures *t* of both junctions. Usually one junction is held at 0°C and then, to a reasonable extent, the following relationship holds:

$$E = at + bt^2$$

where *a* and *b* are constants for the metals concerned. Figure 1.31 shows how the e.m.f. varies with temperature for a number of commonly used pairs of metals. Standard tables giving the e.m.f.s at different temperatures are available for the metals usually used for thermocouples. Commonly used thermocouples are listed in Table 2.1, with the temperature ranges over which they are generally used and typical sensitivities. These commonly used thermocouples are given reference letters. The base-metal thermocouples, E, J, K and T, have accuracies about ±1 to 3%, are relatively cheap but deteriorate with age. Noble-metal thermocouples, e.g. R, have accuracies of about ±1% or better, are more expensive but more stable with longer life. Thermocouples are generally mounted in a sheath to give them mechanical and chemical protection. The response time of an unsheathed thermocouple is very fast. With a sheath this may be increased to as much as a few seconds if a large sheath is used.

Table 1.1 *Thermocouples*

Type	Materials	Range °C	Sensitivity μV/°C
E	Chromel–constantan	0 to 980	63
J	Iron–constantan	−180 to 760	53
K	Chromel–alumel	−180 to 1260	41
R	Platinum–platinum/rhodium 13%	0 to 1750	8
T	Copper–constantan	−180 to 370	43

To maintain one junction of a thermocouple at 0°C, it needs to be immersed in a mixture of ice and water. This, however, is often not convenient. A compensation circuit (Figure 1.32) can, however, be used to provide an e.m.f. which varies with the temperature of the cold junction in such a way that when it is added to the thermocouple e.m.f. it generates a combined e.m.f. which is the same as would have been generated if the cold junction had been at 0°C.

There is a change in the current across the junction of *semiconductor diodes* and *transistors* when the temperature changes. For use as temperature sensors they are supplied, together with the necessary signal processing circuitry, as integrated circuits. An integrated circuit temperature sensor using transistors is LM35. This gives an output, which is a linear function of temperature, of 10 mV/°C when the supply voltage is 5 V. Figure 1.33(a) shows the connections for the range 12°C to 110° and (b) for −40° to 110°.

1.5 Signal processing

The output signal from the sensor of a measurement system or the signal from the control unit might have to be processed in some way to make it suitable to operate the next element in the control system. For example, the signal may be too small and have to be amplified, be analogue and have to be made digital, be digital and have to be made analogue, be a resistance change and have to be made into a current change, be a voltage change and have to be made into a suitable size current change, be a pressure change and have to be made into a current change, etc. All these changes can be referred to as *signal processing*. For example, the output from a thermocouple is a very small voltage, a few millivolts. A signal processing module might then be used to convert this into a larger voltage and provide cold junction compensation (i.e. allow for the cold junction not being at 0°C).

The following are some examples of signal processing commonly encountered in control systems.

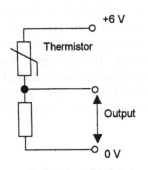

Figure 1.34 *Resistance to voltage conversion for a thermistor*

1.5.1 Resistance to voltage converter

Consider how the resistance change produced by a thermistor when subject to a temperature change can be converted into a voltage change. Figure 1.34 shows how a *potential divider circuit* can be used. A constant voltage, of perhaps 6 V, is applied across the thermistor and another resistor in series. With a thermistor with a resistance of 4.7 kΩ, the series resistor might be 10 kΩ. The output signal is the voltage across the 10 kΩ resistor. When the resistance of the thermistor changes, the fraction of the 6 V across the 10 kΩ resistor changes.

The output voltage is proportional to the fraction of the total resistance which is between the output terminals. Thus:

$$\text{output} = \frac{R}{R + R_t} V$$

where V is the total voltage applied, in Figure 1.34 this is shown as 6 V, R the value of the resistance between the output terminals (10 kΩ) and R_t the resistance of the thermistor at the temperature concerned. The potential divider circuit is thus an example of a simple resistance to voltage converter. Another example of such a converter is the Wheatstone bridge.

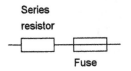

Figure 1.35 *Protection against high currents*

1.5.2 Protection

An important element that is often required with signal processing is protection against high currents or high voltages. A high current can be protected against by the incorporation in the input line of a series resistor to limit the current to an acceptable level and a fuse to break if the current does exceed a safe level (Figure 1.35).

It is often so vital that high currents or high voltages are not transmitted from the sensor to a microprocessor that it may be necessary to completely isolate circuits so there are no electrical connections between them. This can be done using an *optoisolator* (Figure 1.36). Such a device converts an electrical signal into an optical signal,

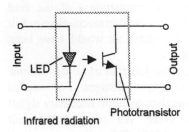

Figure 1.36 *Optoisolator: infrared radiation is used to transmit signal from input to output circuits*

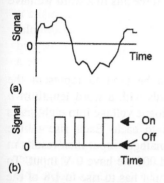

Figure 1.37   *Signals:
(a) analogue, (b) digital*

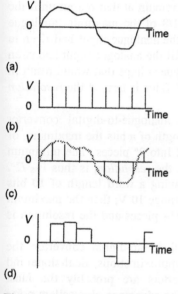

Figure 1.38   *(a) Analogue
signal, (b) time signal, (c)
sampled signal, (d) sampled
and held signal*

transmits it to a detector which then converts it back into an electrical signal. The input signal passes through an infrared light-emitting diode (LED) and so produces a beam of infrared radiation which is detected by a phototransistor.

1.5.3 Analogue to digital conversion

The electrical output from sensors such as thermocouples, resistance elements used for temperature measurement, strain gauges, diaphragm pressure gauges, LVDTs, etc. is in analogue form. An analogue signal (Figure 1.37(a)) is one that is continuously variable, changing smoothly over a range of values. The signal is an analogue, i.e. a scaled version, of the quantity it represents. A digital signal increases in jumps, being a sequence of pulses, often just on-off signals (Figure 1.37(b)). The value of the quantity instead of being represented by the height of the signal, as with analogue, is represented by the sequence of on-off signals.

Microprocessors require digital inputs. Thus, where a microprocessor is used as part of a control system, the analogue output from a sensor has to be converted into a digital form before it can be used as an input to the microprocessor. Thus there is a need for an *analogue-to-digital converter* (ADC). Analogue-to-digital conversion involves a number of stages. The first stage is to take samples of the analogue signal (Figure 1.38(a)). A clock supplies regular time signal pulses (Figure 1.38(b)) to the analogue-to-digital converter and every time it receives a pulse it samples the analogue signal. The result is a series of narrow pulses with heights which vary in accord with the variation of the analogue signal (Figure 1.38(c)). This sequence of pulses is changed into the signal form shown in Figure 1.38(d) by each sampled value being held until the next pulse occurs. This holding is necessary to allow time for the conversion to take place at an analogue-to-digital converter. This converts each sample into a sequence of pulses representing the value. For example, the first sampled value might be represented by 101, the next sample by 011, etc. The 1 represents an 'on' or 'high' signal, the 0 an 'off' or 'low' signal. Analogue-to-digital conversion thus involves a sample and hold unit followed by an analogue- to-digital converter (Figure 1.39).

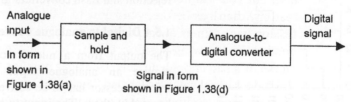

Figure 1.39 *Analogue-to-digital conversion*

To illustrate the action of the analogue-to-digital converter, consider one that gives an output restricted to three bits. The binary digits of 0 and 1, i.e. the 'low' and 'high' signals, are referred to as *bits*. A group of bits is called a *word*. Thus the three bits give the *word length* for this particular analogue-to-digital converter. The word is what represents the

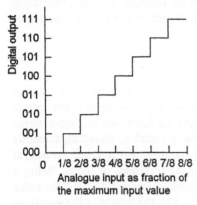

Figure 1.40 *Digital output from an ADC*

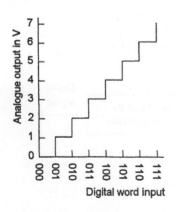

Figure 1.41 *Digital-to-analogue conversion*

digital version of the analogue voltage. With three bits in a word we have the possible words of:

000 001 010 011 100 101 110 111

There are eight possible words which can be used to represent the analogue input; the number of possible words with a word length of n bits is 2^n. Thus we divide the maximum analogue voltage into eight parts and one of the digital words corresponds to each. Each rise in the analogue voltage of (1/8) of the maximum analogue input then results in a further bit being generated. Thus for word 000 we have 0 V input. To generate the next digital word of 001 the input has to rise to 1/8 of the maximum voltage. To generate the next word of 010 the input has to rise to 2/8 of the maximum voltage. Figure 1.40 illustrates this conversion of the sampled and held input voltage to a digital output.

Thus if we had a sampled analogue input of 8 V, the digital output would be 000 for a 0 V input and would remain at that output until the analogue voltage had risen to 1 V, i.e. 1/8 of the maximum analogue input. It would then remain at 001 until the analogue input had risen to 2 V. This value of 001 would continue until the analogue input had risen to 3 V. The smallest change in the analogue voltage that would result in a change in the digital output is thus 1 V. This is termed the *resolution* of the converter.

The word length possible with an analogue-to-digital converter determines its *resolution*. With a word length of n bits the maximum, or full scale, analogue input V_{FS} is divided into 2^n pieces. The minimum change in input that can be detected, i.e. the *resolution*, is thus $V_{FS}/2^n$. With an analogue-to-digital converter having a word length of 10 bits and the maximum analogue signal input range 10 V, then the maximum analogue voltage is divided into $2^{10} = 1024$ pieces and the resolution is $10/1024 = 9.8$ mV.

There are a number of forms of analogue-to-digital converter; the most commonly used being successive approximations, dual-slope and flash. Successive approximations converters are probably the most widely used; dual slope converters have the advantage of excellent noise rejection and flash converters give the highest conversion rates.

1.5.4 Digital to analogue conversion

The output from a microprocessor is digital. Most control elements require an analogue input and so the digital output from a microprocessor has to be converted into an analogue form before it can be used by them. The input to a digital-to-analogue converter is a binary word and the output its equivalent analogue value. For example, if we have a full scale output of 7 V then a digital input of 000 will give 0 V, 001 give 1 V, ... and 111 the full scale value of 7 V. Figure 1.41 illustrates this.

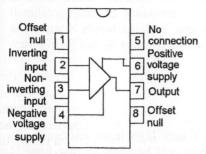

Figure 1.42 *The 741 operational amplifier*

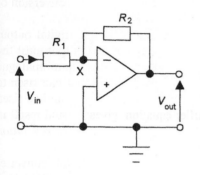

Figure 1.43 *Inverting amplifier*

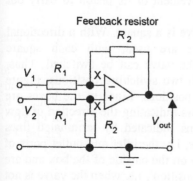

Figure 1.44 *Differential amplifier*

1.5.5 Amplifiers

The *operational amplifier* is the basis of many signal processing elements, the basic amplifier being supplied as an integrated circuit on a silicon chip. It has two inputs, termed the inverting input (−) and the non-inverting input (+) and is a high gain d.c. amplifier, the gain typically being of the order of 100 000 or more. Figure 1.42 shows the pin connections for a 741 operational amplifier with the symbol for the operational amplifier. Pins 4 and 7 are for the connections to the supply voltage for the amplifier, pin 2 for the inverting input, pin 3 for the non-inverting input. The output is taken from pin 6. Pins 1 and 5 are for the offset null. These are to enable circuits to be connected to enable corrections to be made for the non-ideal behaviour of the amplifier.

Consider the amplifier when used as an *inverting amplifier* (Figure 1.43), i.e. an amplifier which gives an output which is out-of-phase with respect to the input. For the circuit shown in Figure 1.43, the connections for the power supply and the offset null have been omitted. The input is connected to the inverting input, the non-inverting input being connected to earth. A feedback loop is connected, via the resistor R_2, to the inverting input. The output voltage of such an amplifier is limited to about ±10 V and thus, since the gain is about 100 000, the input voltage to the inverting input at X, V_X, must be between about +0.0001 V and −0.0001 V. This is virtually zero and so point X is at virtually earth potential. For this reason it is called a *virtual earth*. The potential difference across the input resistance R_1 is $(V_{in} − V_X)$ and thus $(V_{in} − V_X) = I_1R_1$. But V_X is virtually zero and so we can write:

$$V_{in} = I_1R_1$$

Operational amplifiers have very high resistance between their input terminals, e.g. the resistance with the 741 operational amplifier is about 2 MΩ. Thus virtually no current flows from point X through the inverting input and so to earth. Thus the current I_1 that flows through R_1 must be essentially the current flowing through R_2. The potential difference across R_2 is $(V_X − V_{out})$. Thus we can write $(V_X − V_{out}) = I_1R_2$. But as V_X is effectively zero, we can write:

$$−V_{out} = I_2R_2$$

Eliminating I_1 from these two simultaneous equations gives:

$$\text{gain of circuit} = \frac{V_{out}}{V_{in}} = −\frac{R_2}{R_1}$$

The negative sign indicates that the output is 180° out-of-phase with the input. The gain is determined solely by the values of the two resistors. A non-inverting amplifier can likewise be produced by taking the input to the non-inverting input instead of the inverting input.

As an illustration of the use of an operational amplifier, consider Figure 1.44 which shows how it can be used as a differential amplifier to amplify the difference between two input voltages. Since there is virtually no current through the high resistance in the operational amplifier between the two input terminals, both the inputs X will be at the same potential. The voltage V_2 is across resistors R_1 and R_2 in series. Thus the potential V_X at X is:

$$\frac{V_X}{V_2} = \frac{R_2}{R_1 + R_2}$$

Since the operational amplifier has a very high input resistance, the current through the feedback resistance will be equal to that from V_1 through R_1. Hence we have:

$$\frac{V_1 - V_X}{R_1} = \frac{V_X - V_{out}}{R_2}$$

and so:

$$\frac{V_{out}}{R_2} = V_X\left(\frac{1}{R_2} + \frac{1}{R_1}\right) - \frac{V_1}{R_1}$$

Hence substituting for V_X using the earlier equation, gives:

$$V_{out} = \frac{R_2}{R_1}(V_2 - V_1)$$

The output is a measure of the difference between the two input voltages.

1.6 Correction elements

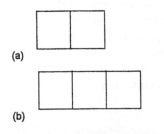

(a)

(b)

Figure 1.45   *(a) Two position, (b) three position valves*

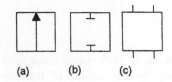

(a)   (b)   (c)

Figure 1.46   *(a) Flow path, (b) shut-off, (c) initial connections*

The following are examples of correction elements that are commonly encountered in control systems.

1.6.1 Directional control valves

A *directional control valve* on the receipt of some external signal, which might be mechanical, electrical or a pressure signal, change the direction of, or stop, or start the flow of fluid in some part of the pneumatic/hydraulic circuit. Thus, it might be used to control the direction of fluid flow to a cylinder and so use the movement of its piston to carry out actuation.

The basic symbol for a control valve is a square. With a directional control valve two or more squares are used, with each square representing the positions to which the valve can be switched. Thus, Figure 1.45(a) represents a valve with two switching positions, Figure 1.45(b) a valve with three switching positions. Lines in the boxes are used to show the flow paths with arrows indicating the direction of flow (Figure 1.46(a)) and shut-off positions indicated by terminated lines (Figure 1.46(b)). The pipe connections, i.e. the inlet and outlet ports of the valve, are indicated by lines drawn on the outside of the box and are drawn for just the 'rest/initial/neutral position', i.e. when the valve is not actuated (Figure 1.46(c)). You can imagine each of the position boxes to be moved by the action of some actuator so that it connects up with the pipe positions to give the different connections between the ports. Directional control valves are described by the number of ports and the number of positions. Thus, a 2/2 valve has 2 ports and 2 positions, a 3/2 valve 3 ports and 2 positions, a 4/2 valve 4 ports and 2 positions, a 5/3 valve 5 ports and 3 positions. Figure 1.47 shows some commonly used examples and their switching options and Figure 1.48 the means by which valves can be switched between positions.

As an illustration, Figure 1.49 shows the symbol for a 3/2 valve with solenoid activation and return by means of a spring. Thus, when the solenoid is not activated by a current through it, the signal port 2 is connected to the exhaust 3 and so is at atmospheric pressure. When the solenoid is activated, the pressure supply P is connected to the signal port 2 and thus the output is pressurised.

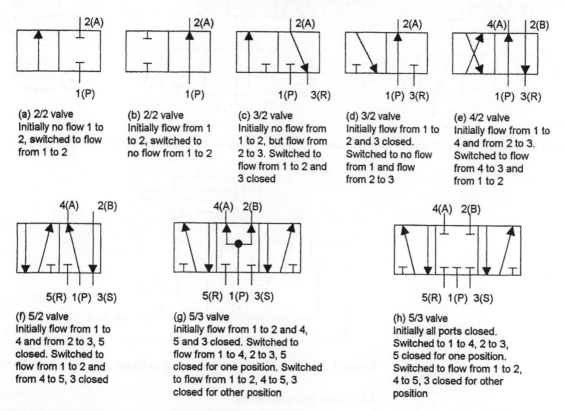

(a) 2/2 valve
Initially no flow 1 to 2, switched to flow from 1 to 2

(b) 2/2 valve
Initially flow from 1 to 2, switched to no flow from 1 to 2

(c) 3/2 valve
Initially no flow from 1 to 2, but flow from 2 to 3. Switched to flow from 1 to 2 and 3 closed

(d) 3/2 valve
Initially flow from 1 to 2 and 3 closed. Switched to no flow from 1 and flow from 2 to 3

(e) 4/2 valve
Initially flow from 1 to 4 and from 2 to 3. Switched to flow from 4 to 3 and from 1 to 2

(f) 5/2 valve
Initially flow from 1 to 4 and from 2 to 3, 5 closed. Switched to flow from 1 to 2 and from 4 to 5, 3 closed

(g) 5/3 valve
Initially flow from 1 to 2 and 4, 5 and 3 closed. Switched to flow from 1 to 4, 2 to 3, 5 closed for one position. Switched to flow from 1 to 2, 4 to 5, 3 closed for other position

(h) 5/3 valve
Initially all ports closed. Switched to 1 to 4, 2 to 3, 5 closed for one position. Switched to flow from 1 to 2, 4 to 5, 3 closed for other position

Figure 1.47 *Commonly used direction valves: P or 1 indicates the pressure supply ports, R and S or 3 and 5 the exhaust ports, A and B or 2 and 4 the signal output ports*

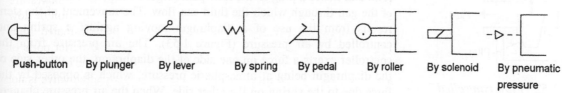

Push-button   By plunger   By lever   By spring   By pedal   By roller   By solenoid   By pneumatic pressure

Figure 1.48 *Examples of valve actuation methods*

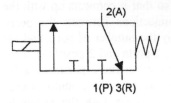

Figure 1.49 *Symbol for a solenoid-activated valve with return spring*

Figure 1.50 shows how such a valve might be used to cause the piston in a single-acting cylinder to move; the term single-acting is used when a pressure signal is applied to only one side of the piston. When the switch is closed and a current passes through the solenoid, the valve switches position and pressure is applied to extend the piston in the cylinder.

Figure 1.51 shows how a double-solenoid activated valve can be used to control a double-acting cylinder. Momentary closing switch S1 causes a current to flow through the solenoid at the left-hand end of the valve and so result in the piston extending. On opening S1 the valve remains in this extended position until a signal is received by the closure of switch S2 to activate the right-hand solenoid and return the piston.

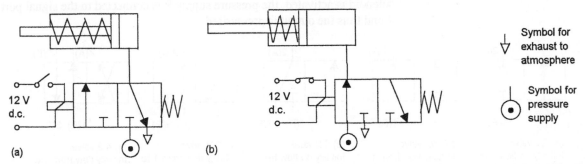

Figure 1.50 *Control of a single-acting cylinder: (a) before solenoid activated, (b) when solenoid activated*

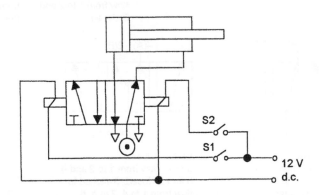

Figure 1.51 *Control of a double-acting cylinder*

1.6.2 Flow control valves

In many control systems the rate of flow of a fluid along a pipe is controlled by a valve which uses pneumatic action to move the valve stem and hence a plug in the flow path (Figure 1.52), so altering the size of the gap through which the fluid can flow. The movement of the stem results from the use of a diaphragm moving against a spring and controlled by air pressure (Figure 1.53). The air pressure from the controller exerts a force on one side of the diaphragm, the other side of the diaphragm being at atmospheric pressure, which is opposed by the force due to the spring on the other side. When the air pressure changes then the diaphragm moves until there is equilibrium between the forces

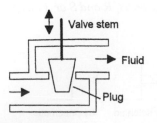

Figure 1.52 *Flow controlled by movement of a plug*

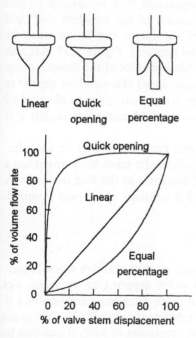

Figure 1.53 *(a) Direct action, (b) reverse action*

Figure 1.55 *Effect of plug shape on flow*

resulting from the pressure and those from the spring. Thus the pressure signals from the controller result in the movement of the stem of the valve. The difference between the direct and reverse forms in Figure 1.53 is the position of the spring.

There are many forms of valve body and plug. The selection of the form of body and plug determine the characteristic of the control valve, i.e. the relationship between the valve stem position and the flow rate through it. For example, Figure 1.54 shows how the selection of plug can be used to determine whether the valve closes when the controller air pressure increases or opens when it increases and Figure 1.55 shows how the shape of the plug determines how the rate of flow is related to the displacement of the valve stem: linear plug – change in flow rate proportional to the change in valve stem displacement; quick-opening plug – a large change in flow rate occurs for a small movement of the valve stem; equal percentage plug – the amount by which the flow rate changes is proportional to the value of the flow rate when the change occurs.

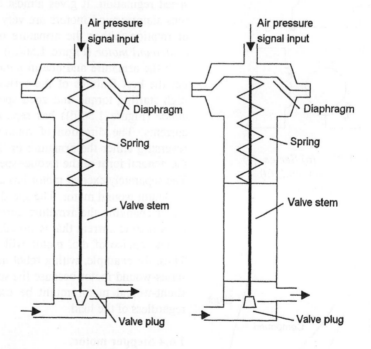

Figure 1.54 *Direct action: (a) air pressure increase to close, (b) air pressure increase to open*

1.6.3 D.c. motors

D.c. motors are widely used with control systems. In the d.c. motor, coils of wire are mounted in slots on a cylinder of magnetic material called the *armature*. The armature is mounted on bearings and is free to rotate. It is mounted in the magnetic field produced by *field poles*. This magnetic field might be produced by permanent magnets or an electromagnet with its magnetism produced by a current passing through the, so-termed,

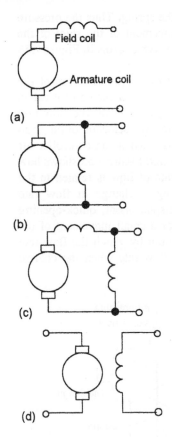

(a)

(b)

(c)

(d)

Figure 1.56   (a) Series, (b) shunt, (c) compound, (d) separately wound

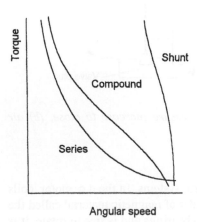

Figure 1.57   Torque–speed characteristics of d.c. motors

field coils. Whether permanent magnet or electromagnet, these generally form the outer casing of the motor and are termed the *stator.*

For a d.c. motor with the field provided by a permanent magnet, the direction of rotation of the motor can be changed by reversing the current in the armature coil. The speed of rotation of such a motor can be changed by changing the size of the current to the armature coil.

D.c. motors with field coils are classified as series, shunt, compound and separately excited according to how the field windings and armature windings are connected. With the *series-wound motor* the armature and fields coils are in series (Figure 1.56(a)). Such a motor exerts the highest starting torque and has the greatest no-load speed. However, with light loads there is a danger that a series-wound motor might run at too high a speed. Reversing the polarity of the supply to the coils has no effect on the direction of rotation of the motor, since both the current in the armature and the field coils are reversed. With the *shunt-wound motor* (Figure 1.56(b)) the armature and field coils are in parallel. It provides the lowest starting torque, a much lower no-load speed and has good speed regulation. It gives almost constant speed regardless of load and thus shunt wound motors are very widely used. To reverse the direction of rotation, either the armature or field current can be reversed. The *compound motor* (Figure 1.56(c)) has two field windings, one in series with the armature and one in parallel. Compound-wound motors aim to get the best features of the series and shunt-wound motors, namely a high starting torque and good speed regulation. The *separately excited motor* (Figure 1.56(d)) has separate control of the armature and field currents. The direction of rotation of the motor can be obtained by reversing either the armature or the field current. Figure 1.57 indicates the general form of the torque–speed characteristics of the above motors. The separately excited motor has a torque–speed characteristic similar to the shunt wound motor. The speed of such d.c. motors can be changed by either changing the armature current or the field current. Generally it is the armature current that is varied.

The choice of d.c. motor will depend on what it is to be used for. Thus, for example, with a robot manipulator the robot wrist might use a series-wound motor because the speed decreases as the load increases. A shunt-wound motor might be used if a constant speed was required, regardless of the load.

1.6.4 Stepper motor

The *stepper or stepping motor* produces rotation through equal angles, the so-called *steps*, for each digital pulse supplied to its input. For example, if with such a motor 1 input pulse produces a rotation of 1.8° then 20 input pulses will produce a rotation through 36.0° , 200 input pulses a rotation through one complete revolution of 360°. It can thus be used for accurate angular positioning. By using the motor to drive a continuous belt, the angular rotation of the motor is transformed into linear motion of the belt and so accurate linear positioning can be achieved. Such a motor is used with computer printers, *x-y* plotters, robots, machine tools and a wide variety of instruments for accurate positioning.

There are two basic forms of stepper motor, the *permanent magnet* type with a permanent magnet rotor and the *variable reluctance* type with a soft steel rotor. Both form of stepper motor have a stator with a number of diametrically opposite pairs of poles, each wound with a coil. Figure 1.58 shows the permanent magnet type with two pairs of stator poles. Each pole is activated by a current being passed through the appropriate field winding, the coils being such that opposite poles are produced on opposite coils. The current is supplied from a d.c. source to the windings through switches. With the currents switched through the coils such that the poles are as shown in Figure 1.58, the rotor will move to line up with the next pair of poles and stop there. This would be, for Figure 1.58, an angle of 45°. If the current is then switched so that the polarities are reversed, the rotor will move a step to line up with the next pair of poles, at angle 135° and stop there. The polarities associated with each step are:

Step	Pole 1	Pole 2	Pole 3	Pole 4
1	North	South	North	South
2	South	North	North	South
3	South	North	South	North
4	North	South	South	North
5	Repeat of steps 1 to 4			

There are thus, in this case, four possible rotor positions: 45°, 135°, 225° and 315°. Note that the term *phase* is used for the number of independent windings on the stator.

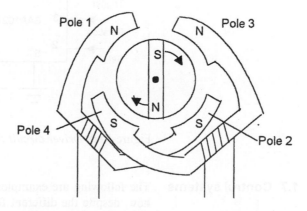

Figure 1.58 *The basic principles of the permanent magnet stepper motor (2-phase) with a rotor giving 90° steps*

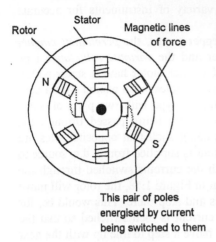

Figure 1.59 *Basic principles of a 3-phase variable reluctance stepper motor*

Figure 1.59 shows the basic form of the *variable reluctance* type of stepper motor. With this form the rotor is made of soft steel and is not a permanent magnet. The rotor has a number of teeth, the number being less than the number of poles on the stator. When an opposite pair of windings on stator poles has current switched to them, a magnetic field is produced with lines of force which pass from the stator poles through the nearest set of teeth on the rotor. Since lines of force can be considered to be rather like elastic thread and always trying to shorten themselves, the rotor will move until the rotor teeth and stator poles line up. This is termed the position of minimum reluctance. Thus by switching the current to successive pairs of stator poles, the rotor can be made to rotate in steps. With the number of poles and rotor teeth shown in Figure 1.59, the angle between each successive step will be 30°. The angle can be made smaller by increasing the number of teeth on the rotor.

To drive a stepper motor, so that it proceeds step-by-step to provide rotation, requires each pair of stator coils to be switched on and off in the required sequence when the input is a sequence of pulses. Driver circuits are available to give the correct sequencing and Figure 1.60 shows an example. The stepper motor will rotate through one step each time the trigger input goes from low to high. The motor runs clockwise when the rotation input is low and anticlockwise when high. When the set pin is made low the output resets. In a control system, these input pulses might be supplied by a microprocessor.

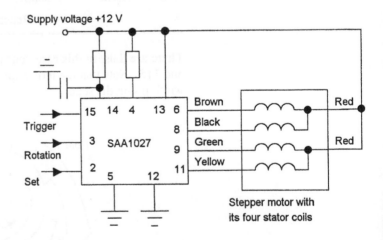

Figure 1.60 *Driver circuit SAA1027 for a 12 V 4-phase stepper motor*

1.7 Control systems

The following are examples of closed-loop control systems to illustrate how, despite the different forms of control being exercised, the systems all have the same basic structural elements.

1.7.1 Control of the speed of rotation of a motor shaft

Consider the motor system shown in Figure 1.61 for the control of the speed of rotation of the motor shaft and its block diagram representation in Figure 1.62. The input of the required speed value is by means of the setting of the position of the movable contact of the potentiometer. This determines what voltage is supplied to the comparison element, i.e. the differential amplifier, as indicative of the required speed of rotation. The differential amplifier produces an amplified output which is proportional to the difference between its two inputs. When there is no difference then the output is zero. The differential amplifier is thus used to both compare and implement the control law. The resulting control signal is then fed to a motor which adjusts the speed of the rotating shaft according to the size of the control signal. The speed of the rotating shaft is measured using a tachogenerator, this being connected to the rotating shaft by means of a pair of bevel gears. The signal from the tachogenerator gives the feedback signal which is then fed back to the differential amplifier.

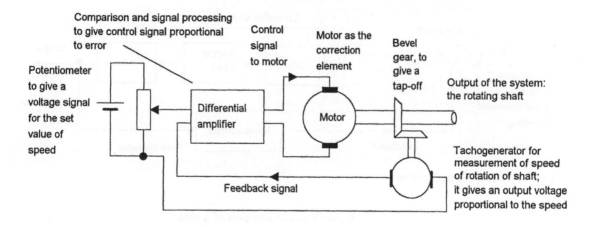

Figure 1.61 *Control of the speed of rotation of a shaft*

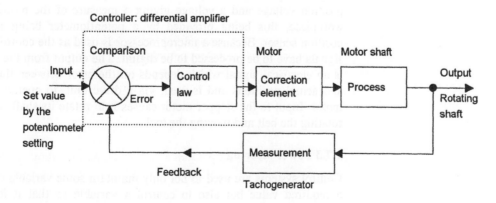

Figure 1.62 *Control of the speed of rotation of a shaft*

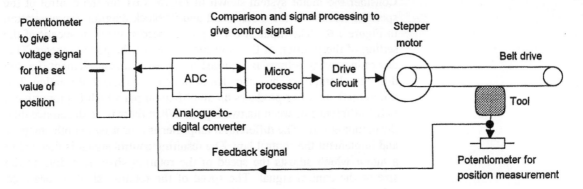

Figure 1.63 *Position control system*

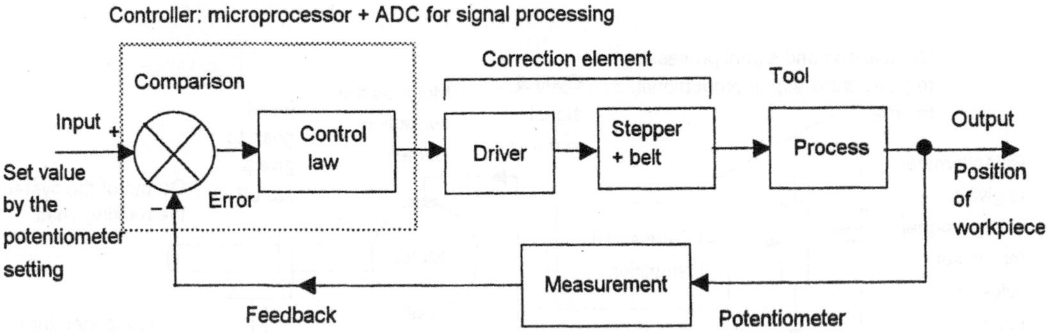

Figure 1.64 *Position control system*

1.7.2 Control of the position of a tool

Figure 1.63 shows a position control system using a belt driven by a stepper motor to control the position of a tool and Figure 1.64 its block diagram representation. The inputs to the controller are the required position voltage and a voltage giving a measure of the position of the workpiece, this being provided by a potentiometer being used as a position sensor. Because a microprocessor is used as the controller, these signals have to be processed to be digital. The output from the controller is an electrical signal which depends on the error between the required and actual positions and is used, via a drive unit, to operate a stepper motor. Input to the stepper motor causes it to rotate its shaft in steps, so rotating the belt and moving the tool.

1.7.3 Power steering

Control systems are used to not only maintain some variable constant at a required value but also to control a variable so that it follows the changes required by a variable input signal. An example of such a control system is the power steering system used with a car. This comes into operation whenever the resistance to turning the steering wheel exceeds a predetermined amount and enables the movement of the wheels to follow the dictates of the angular motion of the steering wheel. The input to the system is the angular position of the steering wheel. This mechanical signal is scaled down by gearing and has subtracted from it a feedback signal representing the actual position of the wheels. This feedback is via a mechanical linkage. Thus when the steering wheel is rotated and there is a difference between its position and the required position of the wheels, there is an error signal. The error signal is used to operate a hydraulic valve and so provide a hydraulic signal to operate a cylinder. The output from the cylinder is then used, via a linkage, to change the position of the wheels. Figure 1.65 shows a block diagram of the system.

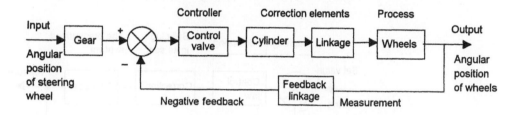

Figure 1.65 *Power assisted steering*

1.7.4 Control of fuel pressure

The modern car involves many control systems. For example, there is the *engine management system* aimed at controlling the amount of fuel injected into each cylinder and the time at which to fire the spark for ignition. Part of such a system is concerned with delivering a constant pressure of fuel to the ignition system. Figure 1.66(a) shows the elements involved in such a system. The fuel from the fuel tank is pumped through a filter to the injectors, the pressure in the fuel line being controlled to be 2.5 bar (2.5×0.1 MPa) above the manifold pressure by a regulator valve. Figure 1.66(b) shows the principles of such a valve. It consists of a diaphragm which presses a ball plug into the flow path of the fuel. The diaphragm has the fuel pressure acting on one side of it and on the other side is the manifold pressure and a spring. If the pressure is too high, the diaphragm moves and opens up the return path to the fuel tank for the excess fuel, so adjusting the fuel pressure to bring it back to the required value.

The pressure control system can be considered to be represented by the closed loop system shown in Figure 1.67. The set value for the pressure is determined by the spring tension. The comparator and control law is given by the diaphragm and spring. The correction element is the ball in its seating and the measurement is given by the diaphragm.

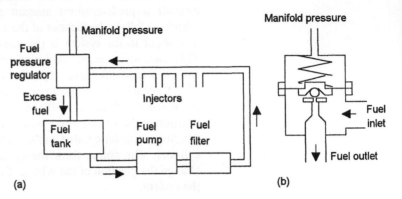

Figure 1.66 *(a) Fuel supply system, (b) fuel pressure regulator*

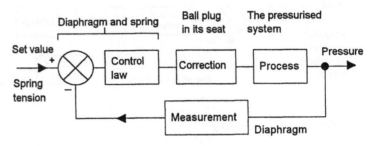

Figure 1.67 *Fuel supply control system*

1.7.5 Antilock brakes

Another example of a control system used with a car is the *antilock brake system (ABS)*. If one or more of the vehicle's wheels lock, i.e. begins to skid, during braking, then braking distance increases, steering control is lost and tyre wear increases. Antilock brakes are designed to eliminate such locking. The system is essentially a control system which adjusts the pressure applied to the brakes so that locking does not occur. This requires continuous monitoring of the wheels and adjustments to the pressure to ensure that, under the conditions prevailing, locking does not occur. Figure 1.68 shows the principles of such a system.

The two valves used to control the pressure are solenoid-operated valves, generally both valves being combined in a component termed the modulator. When the driver presses the brake pedal, a piston moves in a master cylinder and pressurises the hydraulic fluid. This pressure causes the brake calliper to operate and the brakes to be applied. The speed of the wheel is monitored by means of a sensor. When the wheel locks, its speed changes abruptly and so the feedback signal from the sensor changes. This feedback signal is fed into the controller where it is compared with what signal might be expected on the basis of data stored in the controller memory. The controller can then supply output signals which operate the valves and so adjust the pressure applied to the brake.

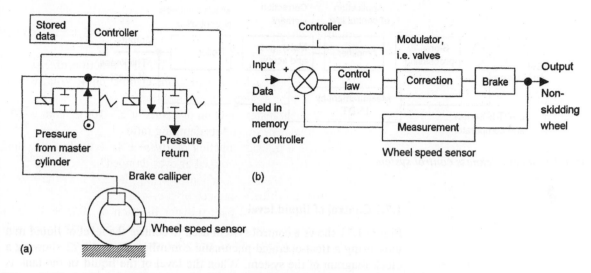

(a)

(b)

Figure 1.68 *Antilock brakes: (a) schematic diagram, (b) block form of the control system*

1.7.6 Thickness control

As an illustration of a process control system, Figure 1.69 shows the type of system that might be used to control the *thickness of sheet* produced by rollers, Figure 1.70 showing the block diagram description of the system. The thickness of the sheet is monitored by a sensor such as a linear variable differential transformer (LVDT). The position of the LVDT probe is set so that when the required thickness sheet is produced, there is no output from the LVDT. The LVDT produces an alternating current output, the amplitude of which is proportional to the error. This is then converted to a d.c. error signal which is fed to an amplifier. The amplified signal is then used to control the speed of a d.c. motor, generally being used to vary the armature current. The rotation of the shaft of the motor is likely to be geared down and then used to rotate a screw which alters the position of the upper roll, hence changing the thickness of the sheet produced.

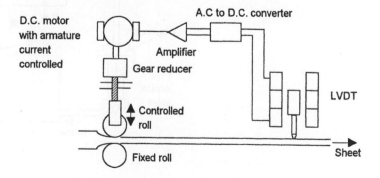

Figure 1.69 *Sheet thickness control system*

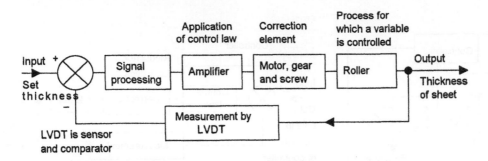

Figure 1.70 *Sheet thickness control system*

1.7.7 Control of liquid level

Figure 1.71 shows a control system used to control the level of liquid in a tank using a float-operated pneumatic controller, Figure 1.72 showing a block diagram of the system. When the level of the liquid in the tank is at the required level and the inflow and outflows are equal, then the controller valves are both closed. If there is a decrease in the outflow of liquid from the tank, the level rises and so the float rises. This causes point P to move upwards. When this happens, the valve connected to the air supply opens and the air pressure in the system increases. This causes a downward movement of the diaphragm in the flow control valve and hence a downward movement of the valve stem and the valve plug. This then results in the inflow of liquid into the tank being reduced. The increase in the air pressure in the controller chamber causes the bellows to become compressed and move that end of the linkage downwards. This eventually closes off the valve so that the flow control valve is held at the new pressure and hence the new flow rate.

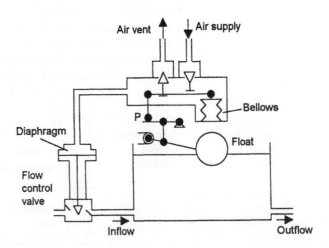

Figure 1.71 *Level control system*

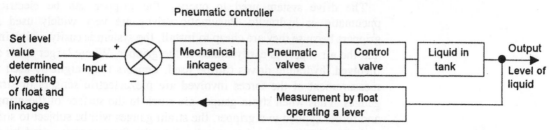

Figure 1.72 *Level control system*

If there is an increase in the outflow of liquid from the tank, the level falls and so the float falls. This causes point P to move downwards. When this happens, the valve connected to the vent opens and the air pressure in the system decreases. This causes an upward movement of the diaphragm in the flow control valve and hence an upward movement of the valve stem and the valve plug. This then results in the inflow of liquid into the tank being increased. The bellows react to this new air pressure by moving its end of the linkage, eventually closing off the exhaust and so holding the air pressure at the new value and the flow control valve at its new flow rate setting.

1.7.8 Robot gripper

The term *robot* is used for a machine which is a reprogrammable multi-function manipulator designed to move tools, parts, materials, etc. through variable programmed motions in order to carry out specified tasks. Here just one aspect will be considered, the gripper used by a robot at the end of its arm to grip objects. A common form of gripper is a device which has 'fingers' or 'jaws'. The gripping action then involves these clamping on the object. Figure 1.73 shows one form such a gripper can take if two gripper fingers are to close on a parallel sided object. When the input rod moves towards the fingers they pivot about their pivots and move closer together. When the rod moves outwards, the fingers move further apart. Such motion needs to be controlled so that the grip exerted by the fingers on an object is just sufficient to grip it, too little grip and the object will fall out of the grasp of the gripper and too great might result in the object being crushed or otherwise deformed. Thus there needs to be feedback of the forces involved at contact between the gripper and the object. Figure 1.74 shows the type of closed-loop control system involved.

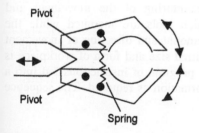

Figure 1.73 *An example of a gripper*

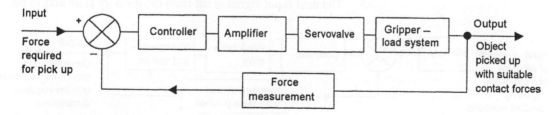

Figure 1.74 *Gripper control system*

The drive system used to operate the gripper can be electrical, pneumatic or hydraulic. Pneumatic drives are very widely used for grippers because they are cheap to install, the system is easily maintained and the air supply is easily linked to the gripper. Where larger loads are involved, hydraulic drives can be used. Sensors that might be used for measurement of the forces involved are piezoelectric sensors or strain gauges. Thus when strain gauges are stuck to the surface of the gripper and forces applied to a gripper, the strain gauges will be subject to strain and give a resistance change related to the forces experienced by the gripper when in contact with the object being picked up.

The robot arm with gripper is also likely to have further control loops to indicate when it is in the right position to grip an object. Thus the gripper might have a control loop to indicate when it is in contact with the object being picked up; the gripper can then be actuated and the force control system can come into operation to control the grasp. The sensor used for such a control loop might be a microswitch which is actuated by a lever, roller or probe coming into contact with the object.

1.7.9 Machine tool control

Machine tool control systems are used to control the position of a tool or workpiece and the operation of the tool during a machining operation. Figure 1.75 shows a block diagram of the basic elements of a closed-loop system involving the continuous monitoring of the movement and position of the work tables on which tools are mounted while the workpiece is being machined. The amount and direction of movement required in order to produce the required size and form of workpiece is the input to the system, this being a program of instructions fed into a memory which then supplies the information as required. The sequence of steps involved is then:

1   An input signal is fed from the memory store.

2   The error between this input and the actual movement and position of the work table is the error signal which is used to apply the correction. This may be an electric motor to control the movement of the work table. The work table then moves to reduce the error so that the actual position equals the required position.

3   The next input signal is fed from the memory store.

4   Step 2 is then repeated.

5   The next input signal is fed from the memory store and so on.

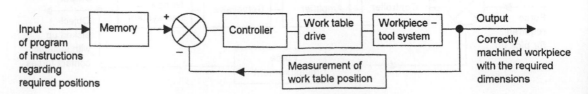

Figure 1.75 *Closed loop machine tool control system*

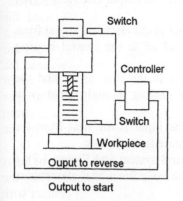

Figure 1.76 *An automatic drill*

1.7.10 An automatic drill

As an illustration of the type of control that might be used with a machine consider the system for a drill which is required to automatically drill a hole in a workpiece when it is placed on the work table (Figure 1.76). A switch sensor can be used to detect when the workpiece is on the work table. This then gives an on input signal to the controller and it then gives an output signal to actuate a motor to lower the drill head and commence drilling. When the drill reaches the full extent of its movement in the workpiece, the drill head triggers another switch sensor. This provides an on input to the controller and it then reverse the direction of rotation of the drill head motor and the drill retracts. This is an example of closed-loop discrete-event control

1.7.11 Microprocessor-controlled systems

The control system used in many modern consumer products, e.g. in a modern motor car or a modern washing machine, to exercise control is likely to be a microprocessor-based system. The controller is then basically as shown in Figure 1.77. It compares the input from a sensor with what is required and then, using a control law determined by the program stored in its memory, gives an output to a correction element.

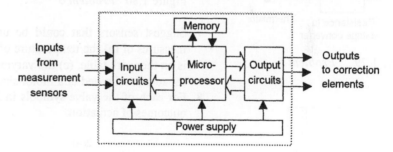

Figure 1.77 *Microprocessor-based controller*

Problems

1 Consider the following as systems and indicate the input and output from each when considered as an entity: (a) a calculator, (b) a loudspeaker, (c) a radio.

2 Explain the difference between open- and closed-loop control systems.

3 Identify the basic functional elements that might be used in the closed-loop control systems involved in:
   (a) A temperature-controlled water bath.
   (b) A speed-controlled electric motor.
   (c) Rollers in a steel strip mill being used to maintain a constant thickness of strip steel.

4 Draw a block diagram of a domestic central heating system which has the following elements:
   (a) A thermostat which has a dial which is set to the required temperature and has an input of the actual temperature in the house

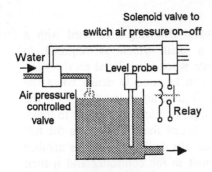

Figure 1.78   *Problem 5*

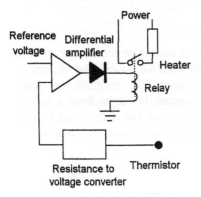

Figure 1.79   *Problem 5*

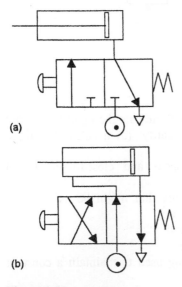

Figure 1.82   *Problem 9*

and which operates as a switch and gives an output electrical signal which is either on or off .

(b) A solenoid valve which has the input of the electrical signal from the thermostat and controls the flow of oil to the central heating furnace.

(c) A heating furnace where the input is the flow of oil and the output is heat to the rooms via water flowing through radiators in the house.

(d) The rooms in the house where the input is the heat from the radiators and the output is the temperature in the rooms.

5   Figure 1.78 shows a temperature control system and Figure 1.79 a water level control system. Identify the basic functional elements of the systems.

6   What will be the relationship between the output and the inputs for the summing elements represented in Figure 1.80?

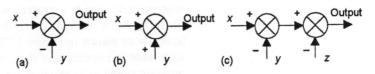

Figure 1.80   *Problem 6*

7   Suggest sensors that could be used with control systems to give measures of (a) the temperature of a liquid, (b) whether a workpiece is on the work table, (c) the varying thickness of a sheet of metal, (d) the rotational speed of a motor shaft.

8   For each of the valve symbols in Figure 1.81, state the method and outcomes of actuation.

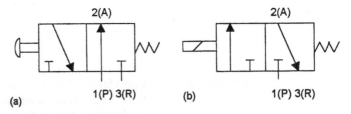

Figure 1.81 *Problem 8*

9   State the outcomes of the pressing and then releasing of the push-buttons with regard to valves shown in Figure 1.82.

10 Draw a block diagram for a negative-feedback system that might be used to control the level of light in a room to a constant value.

11 Draw block diagrams which can be used to present the operation of a toaster when it is (a) an open-loop system, (b) a closed-loop system.

12 Explain how a ball valve is used to control the level of water in a cistern.

2 System models

2.1 Introduction

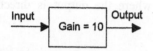

Figure 2.1 *Amplifier system with the output ten times the input*

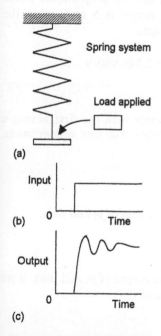

Figure 2.2 *(a) The spring system with a constant load applied at some instant of time, (b) the step showing how the input varies with time, (c) the output showing how it varies with time for the step input*

Suppose we have a control system for the temperature in a room. How will the temperature react when the thermostat has its set value increased from, say, 20°C to 22°C? In order to determine how the output of a control system will react to different inputs, we need a mathematical model of the system so that we have an equation describing how the output of the system is related to its input.

Thus, in the case of an amplifier system (Figure 2.1) we might be able to use the simple relationship that the output is always 10 times the input. If we have an input of a 1 V signal we can calculate that the output will be 10 V. This is a simple model of a system where the input is just multiplied by a gain of 10 in order to give the output. This chapter starts off with a discussion of this simple model of a system.

However, if we consider a system representing a spring balance with an input of a load signal and an output of a deflection (Figure 2.2) then, when we have an input to the system and put a fixed load on the balance (this type of input is known as a *step input* because the input variation with time looks like a step), it is likely that it will not instantaneously give the weight but the pointer on the spring balance will oscillate for a little time before settling down to the weight value. Thus we cannot just state, for an input of some constant load, that the output is just the input multiplied by some constant number but need some way of describing an output which varies with time. With an electrical system of a circuit with capacitance and resistance, when the voltage to such a circuit is switched on, i.e. there is a constant voltage input to the system, then the current changes with time before eventually settling down to a steady value. With a temperature control system, such as that used for the central heating system for a house, when the thermostat is changed from 20°C to 22°C, the output does not immediately become 22°C but there is a change with time and eventually it may become 22°C. In general, the mathematical model describing the relationship between input and output for a system is likely to involve terms which give values which change with time and are described by a differential equation (see Appendix A). In this chapter we look at how such differential equation relationships arise.

In order to make life simple, what we need is a simple relationship between input and output for a system, even when the output varies with time. It is nice and simple to say that the output is just ten times the input and so describe the system by gain = 10. There is a way we can have such a simple form of relationship where the relationship involves time but it involves writing inputs and outputs in a different form. It is called the *Laplace transform*. In this chapter we will consider how we can carry out such transformations, but not the mathematics to justify why we can do it; the aim is to enable you to use the transform as a tool to carry out tasks. Appendix B gives an explanation of the mathematics behind the transform and the way it is used.

2.2 Gain

In the case of an amplifier system we might have the output directly proportional to the input and, with a gain of 10, if we have an input of a 1 V signal we can calculate that the output will be ten times greater and so 10 V. In general, for such a system where the output is directly proportional to the input, we can write:

output = G × input

with G being the gain.

Example

A motor has an output speed which is directly proportional to the voltage applied to its armature. If the output is 5 rev/s when the input voltage is 2 V, what is the system gain?

With output = G × input, then $G = 5/2 = 2.5$ (rev/s)/V.

2.2.1 Gain of systems in series

Consider two systems, e.g. amplifiers, in series with the first having a gain G_1 and the second a gain G_2 (Figure 2.3(a)). The first system has an input of x_1 and an output of y_2 and thus:

$y_1 = G_1 x_1$

The second system has an input of y_1 and an output of y_2 and thus:

$y_2 = G_2 y_1 = G_2 × G_1 x_1$

The overall system has an input of x_1 and an output of y_2 and thus, if we represent the overall system as having a gain of G:

$y_2 = G x_1$

and so:

$G = G_1 × G_2$

Thus:

For series-connected systems, the overall gain is the product of the gains of the constituent systems.

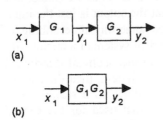

(a)

(b)

Figure 2.3   *(a) Two systems in series, (b) the equivalent system with a gain equal to the product of the gains of the two constituent systems*

Example

A system consists of an amplifier with a gain of 10 providing the armature voltage for a motor which gives an output speed which is proportional to the armature voltage, the constant of proportionality being 5 (rev/s)/V. What is the relationship between the input voltage to the system and the output motor speed?

The overall gain $G = G_1 \times G_2 = 10 \times 5 = 50$ (rev/s)/V.

2.2.2 Feedback loops

Consider a system with negative feedback (Figure 2.4). The output of the system is fed back via a measurement system with a gain H to subtract from the input to a system with gain G.

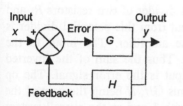

Figure 2.4   *System with negative feedback: the fed back signal subtracts from the input*

The input to the feedback system is y and thus its output, i.e. the feedback signal, is Hy. The error is $x - Hy$. Hence, the input to the G system is $x - Hy$ and its output y. Thus:

$$y = G(x - Hy)$$

and so:

$$(1 + GH)y = Gx$$

The overall input of the system is y for an input x and so the overall gain G of the system is y/x. Hence:

$$\text{system gain} = \frac{y}{x} = \frac{G}{1 + GH}$$

For a system with a negative feedback, the overall gain is the forward path gain divided by one plus the product of the forward path and feedback path gains.

For a system with positive feedback (Figure 2.5), i.e. the fed back signal adds to the input signal, the feedback signal is Hy and thus the input to the G system is $x + Hy$. Hence:

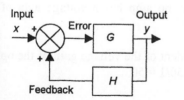

Figure 2.4   *System with positive feedback: the fed back signal adds to the input*

$$y = G(x + Hy)$$

and so:

$$(1 - GH)y = Gx$$

$$\text{system gain} = \frac{y}{x} = \frac{G}{1 - GH}$$

For a system with a positive feedback, the overall gain is the forward path gain divided by one minus the product of the forward path and feedback path gains.

Example

A negative feedback system has a forward path gain of 12 and a feedback path gain of 0.1. What is the overall gain of the system?

$$\text{System gain} = \frac{G}{1+GH} = \frac{12}{1+0.1\times 12} = 5.45$$

2.2.3 The feedback amplifier

Figure 2.6 shows the circuit of a basic feedback amplifier. It consists of an operational amplifier with a potential divider of two resistors R_1 and R_2 connected across its output. The output from this potential divider is fed back to the inverting input of the amplifier. The input to the amplifier is via its non-inverting input. Thus the sum of the inverted feedback input and the non-inverted input is the error signal. The op amp has a very high voltage gain G. Thus GH, H being the gain of the feedback loop, is very large compared with 1 and so the overall system gain is:

$$\text{system gain} = \frac{G}{1+GH} \simeq \frac{G}{GH} \simeq \frac{1}{H}$$

Since the gain G of the op amp can be affected by changes in temperature, ageing, etc. and thus can vary, the use of the op amp with a feedback loop means that, since H is just made up of resistances which are likely to be more stable, a more stable amplifier system is produced. The feedback loop gain H is the fraction of the output signal fed back and so is $R_1/(R_1 + R_2)$. Hence, the overall gain of the system is:

$$\text{system gain} = \frac{R_2+R_1}{R_1}$$

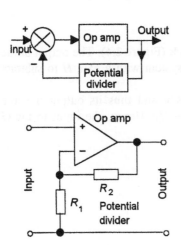

Figure 2.6   *Feedback amplifier*

Example

What is the overall gain of a non-inverting feedback op amp, connected as in Figure 2.6, if the op amp has a voltage gain of 200 000, $R_1 = 1$ kΩ and $R_2 = 49$ kΩ?

The overall system gain is independent of the voltage gain of the op amp and is given by $(R_1 + R_2)/R_1 = 50/1 = 50$.

2.3 Dynamic systems

The following describes how we can arrive at the input–output relationships for systems by representing them by simple models obtained by considering them to be composed of just a few simple basic elements.

2.3.1 Mechanical systems

Mechanical systems, however complex, have stiffness (or springiness), damping and inertia and can be considered to be composed of basic elements which can be represented by springs, dashpots and masses.

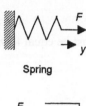

Spring

Dashpot

Mass

Figure 2.7   *Mechanical system building blocks*

1   Spring

The 'springiness' or 'stiffness' of a system can be represented by a spring. For a linear spring (Figure 2.7(a)),   the extension y is proportional to the applied extending force F and we have:

$$F = ky$$

where k is a constant termed the *stiffness*.

2   Dash pot

The 'damping' of a mechanical system can be represented by a dashpot. This is a piston moving in a viscous medium in a cylinder (Figure 2.7(b)). Movement of the piston inwards requires the trapped fluid to flow out past edges of the piston; movement outwards requires fluid to flow past the piston and into the enclosed space. The resistive force F which has to be overcome is proportional to the velocity of the piston and hence the rate of change of displacement y with time, i.e. dy/dt. Thus:

$$F = c\frac{dy}{dt}$$

where c is a constant.

3   Mass

The 'inertia' of a system, i.e. how much it resists being accelerated can be represented by mass. For a mass m (Figure 2.7(c)), the relationship between the applied force F and its acceleration a is given by Newton's second law as $F = ma$. But acceleration is the rate of change of velocity v with time t, i.e. $a = dv/dt$, and velocity is the rate of change of displacement y with time, i.e. $v = dy/dt$. Thus $a = d(dy/dt)/dt$ and so we can write:

$$F = m\frac{d^2y}{dt^2}$$

The following example illustrates how we can arrive at a model for a mechanical system.

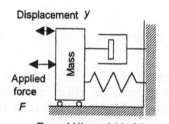

Displacement y

Mass

Applied force

F

To avoid the weight of the mass becoming involved, the forces and displacement are horizontal. The rollers enable us to ignore friction.

(a)

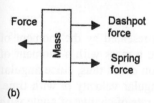

Force

Mass

Dashpot force

Spring force

(b)

Figure 2.8   *(a) Mechanical system with mass, damping and stiffness, (b) the free-body diagram for the forces acting on the mass*

Example

Derive a model for the mechanical system given in Figure 2.8(a). The input to the system is the force F and the output is the displacement y.

To obtain the system model we draw *free-body diagrams*, these being diagrams of masses showing just the external forces acting on each mass. For the system in Figure 2.8(a) we have just one mass and so just one free-body diagram and that is shown in Figure 2.8(b). As the free-body diagram indicates, the net force acting on the mass is the applied force minus the forces exerted by the spring and by the dashpot:

$$\text{net force} = F - ky - c\frac{\mathrm{d}y}{\mathrm{d}t}$$

Then applying Newton's second law, this force must be equal to ma, where a is the acceleration, and so:

$$m\frac{\mathrm{d}^2y}{\mathrm{d}t^2} = F - ky - c\frac{\mathrm{d}y}{\mathrm{d}t}$$

The relationship between the input F to the system and the output y is thus described by the second-order differential equation:

$$m\frac{\mathrm{d}^2y}{\mathrm{d}t^2} + c\frac{\mathrm{d}y}{\mathrm{d}t} + ky = F$$

The term *second-order* is used because the equation includes as its highest derivative $\mathrm{d}^2y/\mathrm{d}t^2$.

2.3.2 Rotational systems

For rotational systems the basic building blocks are a torsion spring, a rotary damper and the moment of inertia (Figure 2.9).

1   *Torsional spring*
    The 'springiness' or 'stiffness' of a rotational spring is represented by a torsional spring. For a torsional spring, the angle θ rotated is proportional to the torque T:

$$T = k\theta$$

where k is a measure of the stiffness of the spring.

2   *Rotational dashpot*
    The damping inherent in rotational motion is represented by a rotational dashpot. For a rotational dashpot, i.e. effectively a disk rotating in a fluid, the resistive torque T is proportional to the angular velocity ω and thus:

$$T = c\omega = c\frac{\mathrm{d}\theta}{\mathrm{d}t}$$

where c is the damping constant.

3   *Inertia*
    The inertia of a rotational system is represented by the moment of inertia of a mass. A torque T applied to a mass with a moment of inertia I results in an angular acceleration a and thus, since angular acceleration is the rate of change of angular velocity ω with time, i.e. $\mathrm{d}\omega/\mathrm{d}t$, and angular velocity ω is the rate of change of angle with time, i.e. $\mathrm{d}\theta/\mathrm{d}t$, then the angular acceleration is $\mathrm{d}(\mathrm{d}\theta/\mathrm{d}t)/\mathrm{d}t$ and so:

$$T = Ia = I\frac{\mathrm{d}^2\theta}{\mathrm{d}t^2}$$

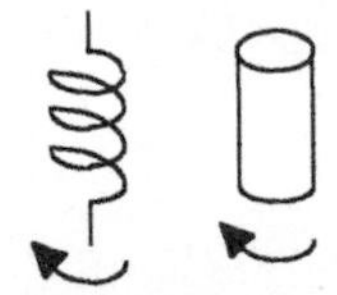

(a) Torsional spring or elastic twisting of a shaft

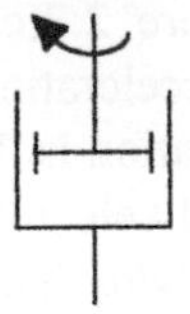

(b) Rotational dashpot

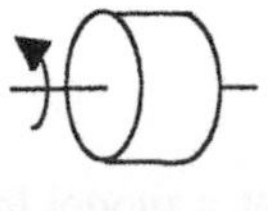

(c) Moment of inertia

Figure 2.9 *Rotational system elements: (a) torsional spring, (b) rotational dashpot, (c) moment of inertia*

The following example illustrates how we can arrive at a model for a rotational system.

Example

Develop a model for the system shown in Figure 2.10(a) of the rotation of a disk as a result of twisting a shaft.

Figure 2.10(b) shows the free-body diagram for the system. The torques acting on the disk are the applied torque T, the spring torque $k\theta$ and the damping torque $c\omega$. Hence:

$$T - k\theta - c\frac{\mathrm{d}\theta}{\mathrm{d}t} = I\frac{\mathrm{d}^2\theta}{\mathrm{d}t^2}$$

We thus have the second-order differential equation relating the input of the torque to the output of the angle of twist:

$$I\frac{\mathrm{d}^2\theta}{\mathrm{d}t^2} + c\frac{\mathrm{d}\theta}{\mathrm{d}t} + k\theta = T$$

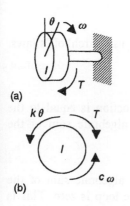

(a)

(b)

Figure 2.10   *Example*

2.3.3 Electrical systems

The basic elements of electrical systems are the resistor, inductor and capacitor (Figure 2.11).

1   *Resistor*

For a *resistor*, resistance R, the potential difference v across it when there is a current i through it is given by:

$$v = Ri$$

2   *Inductor*

For an *inductor*, inductance L, the potential difference v across it at any instant depends on the rate of change of current i and is:

$$v = L\frac{\mathrm{d}i}{\mathrm{d}t}$$

3   *Capacitor*

For a *capacitor*, the potential difference v across it depends on the charge q on the capacitor plates with $v = q/C$, where C is the capacitance. Thus:

$$v = \frac{1}{C}q$$

$$\frac{\mathrm{d}v}{\mathrm{d}t} = \frac{1}{C}\frac{\mathrm{d}q}{\mathrm{d}t}$$

Since current i is the rate of movement of charge:

Resistor

Inductor

Capacitor

Figure 2.11   *Electrical system building blocks*

$$\frac{dv}{dt} = \frac{1}{C}\frac{dq}{dt} = \frac{1}{C}i$$

and so we can write:

$$i = C\frac{dv}{dt}$$

To develop the models for electrical circuits we use Kirchhoff's laws. These can be stated as:

1   *Kirchhoff's current law*
The total current flowing into any circuit junction is equal to the total current leaving that junction, i.e. the algebraic sum of the currents at a junction is zero.

2   *Kirchhoff's voltage law*
In a closed circuit path, termed a loop, the algebraic sum of the voltages across the elements that make up the loop is zero. This is the same as saying that for a loop containing a source of e.m.f., the sum of the potential drops across each circuit element is equal to the sum of the applied e.m.f.'s. provided we take account of their directions.

The following examples illustrate the development of models for electrical systems.

Example

Develop a model for the electrical system described by the circuit shown in Figure 2.12. The input is the voltage v when the switch is closed and the output is the voltage v_C across the capacitor.

Using Kirchhoff's voltage law gives:

$$v = v_R + v_C$$

and, since $v_R = Ri$ and $i = C(dv_C/dt)$ we obtain the equation:

$$v = RC\frac{dv_C}{dt} + v_C$$

The relationship between an input v and the output v_C is a first order differential equation. The term *first-order* is used because it includes as its highest derivative dv_C/dt.

Example

Develop a model for the circuit shown in Figure 2.13 when we have an input voltage v when the switch is closed and take an output as the voltage v_C across the capacitor.

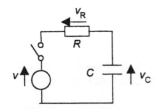

Figure 2.12   *Electrical system with resistance and capacitance*

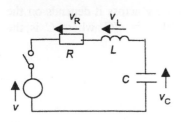

Figure 2.13   *Electrical system with resistance, inductance and capacitance*

Applying Kirchhoff's voltage law gives:

$$v = v_R + v_L + v_C$$

and so:

$$v = Ri + L\frac{di}{dt} + v_C$$

Since $i = C(dv_C/dt)$, then $di/dt = C(d^2v_C/dt^2)$ and thus we can write:

$$v = RC\frac{dv_C}{dt} + LC\frac{d^2v_C}{dt^2} + v_C$$

The relationship between an input v and output v_C is described by a second order differential equation.

2.3.4 Thermal systems

Thermal systems have two basic building blocks, resistance and capacitance (Figure 2.14).

1 *Thermal resistance*
The thermal resistance R is the resistance offered to the rate of flow of heat q (Figure 2.14(a)) and is defined by:

$$q = \frac{T_1 - T_2}{R}$$

where $T_1 - T_2$ is the temperature difference through which the heat flows.

For heat conduction through a solid we have the rate of flow of heat proportional to the cross-sectional area and the temperature gradient. Thus for two points at temperatures T_1 and T_2 and a distance L apart:

$$q = Ak\frac{T_1 - T_2}{L}$$

with k being the thermal conductivity. Thus with this mode of heat transfer, the thermal resistance R is L/Ak. For heat transfer by convection between two points, Newton's law of cooling gives:

$$q = Ah(T_2 - T_1)$$

where $(T_2 - T_1)$ is the temperature difference, h the coefficient of heat transfer and A the surface area across which the temperature difference is. The thermal resistance with this mode of heat transfer is thus $1/Ah$.

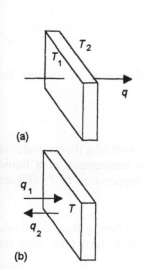

(a)

(b)

Figure 2.14   *(a) Thermal resistance, (b) thermal capacitance*

2   *Thermal capacitance*

The thermal capacitance (Figure 2.14(b)) is a measure of the store of internal energy in a system. If the rate of flow of heat into a system is q_1 and the rate of flow out q_2 then the rate of change of internal energy of the system is $q_1 - q_2$. An increase in internal energy can result in a change in temperature:

change in internal energy = mc × change in temperature

where m is the mass and c the specific heat capacity. Thus the rate of change of internal energy is equal to mc times the rate of change of temperature. Hence:

$$q_1 - q_2 = mc\frac{dT}{dt}$$

This equation can be written as:

$$q_1 - q_2 = C\frac{dT}{dt}$$

where the capacitance $C = mc$.

The following examples illustrates the development of models for thermal systems.

Example

Develop a model for the simple thermal system of a thermometer at temperature T being used to measure the temperature of a liquid when it suddenly changes to the higher temperature of T_L (Figure 2.15).

When the temperature changes there is heat flow q from the liquid to the thermometer. The thermal resistance to heat flow from the liquid to the thermometer is:

$$q = \frac{T_L - T}{R}$$

Since there is only a net flow of heat from the liquid to the thermometer the thermal capacitance of the thermometer is:

$$q = C\frac{dT}{dt}$$

Substituting for q gives:

$$C\frac{dT}{dt} = \frac{T_L - T}{R}$$

which, when rearranged gives:

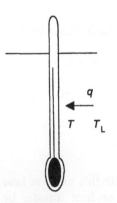

Figure 2.15   *Example*

$$RC\frac{dT}{dt} + T = T_L$$

This is a first-order differential equation.

Example

Determine a model for the temperature of a room (Figure 2.16) containing a heater which supplies heat at the rate q_1 and the room loses heat at the rate q_2.

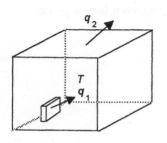

Figure 2.16 *Example*

We will assume that the air in the room is at a uniform temperature T. If the air and furniture in the room have a combined thermal capacity C, since the energy rate to heat the room is $q_1 - q_2$, we have:

$$q_1 - q_2 = C\frac{dT}{dt}$$

If the temperature inside the room is T and that outside the room T_0 then

$$q_2 = \frac{T - T_0}{R}$$

where R is the thermal resistance of the walls. Substituting for q_2 gives:

$$q_1 - \frac{T - T_0}{R} = C\frac{dT}{dt}$$

Hence:

$$RC\frac{dT}{dt} + T = Rq_1 + T_0$$

This is a first-order differential equation.

2.3.5 Hydraulic systems

For a fluid system the three building blocks are resistance, capacitance and inertance; these are the equivalents of electrical resistance, capacitance and inductance. The equivalent of electrical current is the volumetric rate of flow and of potential difference is pressure difference. Hydraulic fluid systems are assumed to involve an incompressible liquid; pneumatic systems, however, involve compressible gases and consequently there will be density changes when the pressure changes. Here we will just consider the simpler case of hydraulic systems. Figure 2.17 shows the basic form of building blocks for hydraulic systems.

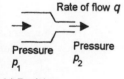

(a) Resistance

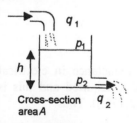

(b) Capacitance

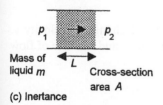

(c) Inertance

Figure 2.17 *Hydraulic building blocks*

1 *Hydraulic resistance*

Hydraulic resistance R is the resistance to flow which occurs when a liquid flows from one diameter pipe to another (Figure 2.17(a)) and is defined as being given by the hydraulic equivalent of Ohm's law:

$$p_1 - p_2 = Rq$$

2 *Hydraulic capacitance*

Hydraulic capacitance C is the term used to describe energy storage where the hydraulic liquid is stored in the form of potential energy (Figure 2.17(b)). The rate of change of volume V of liquid stored is equal to the difference between the volumetric rate at which liquid enters the container q_1 and the rate at which it leaves q_2, i.e.

$$q_1 - q_2 = \frac{dV}{dt}$$

But $V = Ah$ and so:

$$q_1 - q_2 = A\frac{dh}{dt}$$

The pressure difference between the input and output is:

$$p_1 - p_2 = p = h\rho g$$

Hence, substituting for h gives:

$$q_1 - q_2 = \frac{A}{\rho g}\frac{dp}{dt}$$

The hydraulic capacitance C is defined as:

$$C = \frac{A}{\rho g}$$

and thus we can write:

$$q_1 - q_2 = C\frac{dp}{dt}$$

3 *Hydraulic inertance*

Hydraulic inertance is the equivalent of inductance in electrical systems. To accelerate a fluid a net force is required and this is provided by the pressure difference (Figure 2.17(c)). Thus:

$$(p_1 - p_2)A = ma = m\frac{dv}{dt}$$

where a is the acceleration and so the rate of change of velocity v. The mass of fluid being accelerated is $m = AL\rho$ and the rate of flow $q = Av$ and so:

$$(p_1 - p_2)A = L\rho\frac{dq}{dt}$$

$$p_1 - p_2 = I\frac{dq}{dt}$$

where the inertance I is given by $I = L\rho/A$.

The following example illustrates the development of a model for a hydraulic system.

Example

Develop a model for the hydraulic system shown in Figure 2.18 where there is a liquid entering a container at one rate q_1 and leaving through a valve at another rate q_2.

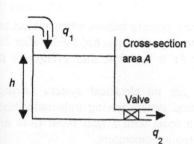

Figure 2.18 *Example*

We can neglect the inertance since flow rates can be assumed to change only very slowly. For the capacitance term we have:

$$q_1 - q_2 = C\frac{dp}{dt} = \frac{A}{\rho g}\frac{dp}{dt}$$

For the resistance of the valve we have:

$$p_1 - p_2 = Rq_1$$

Thus, substituting for q_2, and recognising that the pressure difference is $h\rho g$, gives:

$$q_1 = A\frac{dh}{dt} + \frac{h\rho g}{R}$$

$$A\frac{dh}{dt} + \frac{\rho g}{R}h = q_1$$

This is a first-order differential equation.

2.4 Differential equations

As the previous section indicates, the relationship between the input and output for systems is often in the form of a differential equation which shows how, when there is some input, the output varies with time. See Appendix A for a discussion of differential equations and how the outputs can be derived, here we consider only the results.

2.4 First-order differential equations

Many systems have input–output relationships which can be described by a first-order differential equation and have an output y related to an input x by an equation of the form:

$$\tau\frac{dy}{dt} + y = kx$$

where τ and k are constants, τ being known as the *time constant*.

Consider the response of such a system when subject to a unit step input, i.e. an input which suddenly changes from 0 to a constant value of 1. When we reach the time at which the input x is not changing with

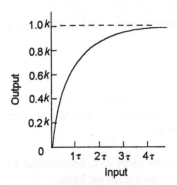

Figure 2.19   *Behaviour of a first order system when subject to a unit step input*

time, i.e. we have steady-state conditions, then $dx/dt = 0$ and so we have output $y = kx$ and k is the *steady-state gain*. Thus, with a unit step input the steady-state output is $1k$. Over time, the output is related to the input by an equation of the form:

$$y = \text{steady-state value} \times (1 - e^{-t/\tau})$$

Thus, Figure 2.19 shows how first order systems behave when subject to a unit step input. After a time of 1τ the output has reached $0.63k$, after 2τ is $0.86k$, after 3τ it is $0.95k$, after 4τ it is $0.98k$, and eventually it becomes $1k$.

Examples of first-order systems are an electrical system having capacitance and resistance, an electrical system having inductance and resistance and a thermal system of a room with a heat input from an electrical heater and an output of the room temperature.

2.4.2 Second-order differential equations

Many systems have input–output relationships which can be described by second-order differential equations and have an output y related to an input x by an equation of the form:

$$\frac{d^2y}{dt^2} + 2\zeta\omega_n\frac{dy}{dt} + \omega_n^2 y = k\omega_n^2 x$$

where k, ζ and ω_n are constants for the systems. The constant ζ is known as the *damping ratio* or *factor* and ω_n as the *undamped natural angular frequency*.

If the input y is not changing with time, i.e. we have steady-state conditions, then $d^2y/dt^2 = 0$ and $dy/dt = 0$ and so we have output $y = kx$ and k is the *steady-state gain*. Figure 2.20 shows how a second order system behaves when subject to a unit step input.

The general form of the response varies with the damping factor. Systems with damping factors less than 1 are said to be *underdamped*, with damping factors greater than 1 as *overdamped* and for a damping factor of 1 as *critically damped*.

1   With no damping, i.e. $\zeta = 0$, the system output oscillates with a constant amplitude and a frequency of ω_n (since $\omega_n = 2\pi f_n$, where f_n is the undamped natural frequency, and $f_n = 1/T$, where T_n is the time for one undamped oscillation, then $T_n = 2\pi/\omega_n = 6.3/\omega_n$).

2   With underdamping i.e. $\zeta < 0$, the output oscillates but the closer the damping factor is to 1 the faster the amplitude of the oscillations diminishes.

3   With critical damping, i.e. $\zeta = 1$, there are no oscillations and the output just gradually approaches the steady-state value.

4   With overdamping, i.e. $\zeta > 1$, the output takes longer than critical damping to reach the steady-state value.

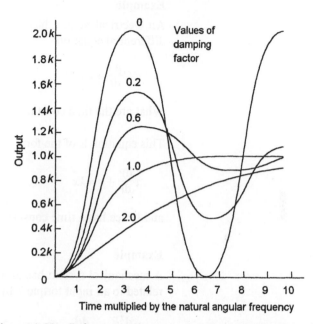

Figure 2.20 *Behaviour of a second order system with a unit step input. With no damping the output is just a continuous oscillation following a step input. As the damping increases, so the oscillations become damped out and with a damping factor of 1.0 there are no oscillations and the output just rises over time to the steady state output value. Further increases in damping mean that the output takes longer to reach the steady state value.*

A mechanical system which can be modelled by a spring, dash pot and mass is an example of a second order system. When we apply a load to the system then oscillations occur which have amplitudes which die away with time. This was illustrated in the opening section and Figure 2.2. Likewise with the second order system of an electrical circuit having resistance, inductance and capacitance; when there is a step voltage input, i.e. a switch is closed and applies a constant voltage to the circuit, then the voltage across the capacitor will be described by a second order differential equation and so can oscillate with amplitudes which die away with time.

The differential equations describe the input/output relationship when we consider the input and output to be functions of time. We can use the model building techniques described in the previous section to arrive at differential equations, alternatively we can find the response of a system to, say, a step input and by examining the response determine the form of the differential equation which described its behaviour. In Chapter 6 we consider the response of systems to sinusoidal inputs and use the response to determine the form of differential equation which describes its behaviour.

Example

An electrical system has an output v related to the input V by the differential equation:

$$RC\frac{dv}{dt} + v = V$$

What are the time constant and the steady state gain of the system?

This equation is of the form:

$$\tau\frac{dy}{dt} + y = kx$$

and hence has a time constant of RC and a steady state gain of 1.

Example

A mechanical system has an output of a rotation through an angle θ related to an input torque T by the differential equation:

$$I\frac{d^2\theta}{dt^2} + c\frac{d\theta}{dt} + k\theta = T$$

What are the natural angular frequency and the damping constant of the system?

The equation can be put into the form:

$$\frac{d^2y}{dt^2} + 2\zeta\omega_n\frac{dy}{dt} + \omega_n^2 y = k\omega_n^2 x$$

as:

$$\frac{d^2\theta}{dt^2} + \frac{c}{I}\frac{d\theta}{dt} + \frac{k}{I}\theta = \frac{1}{I}T$$

Hence, comparing the terms in front of y and θ gives $\omega_n = \sqrt{(k/I)}$. Comparing the terms in front of dy/dt and $d\theta/dt$ gives $2\zeta\omega_n = c/I$ and hence, since $\omega_n = \sqrt{(k/I)}$, gives $\zeta = c/[2\sqrt{(kI)}]$.

2.5 Transfer function

In general, when we consider inputs and outputs of systems as functions of time then the relationship between the output and input is given by a differential equation. If we have a system composed of two elements in series with each having its input–output relationships described by a differential equation, it is not easy to see how the output of the system as a whole is related to its input. There is a way we can overcome this problem and that is to transform the differential equations into a more convenient form by using the *Laplace transform* (see Appendix B for mathematical details). This form is a much more convenient way of describing the relationship than a differential equation since it can be easily manipulated by the basic rules of algebra.

To carry out the transformation we follow the following rules:

1   A variable which is a function of time, e.g. the input or output voltage v in a circuit (to emphasise that v is a function of time it might be written as $v(t)$ – note that this does not mean that v is multiplied by t), becomes a function of s. A voltage is thus written as $V(s)$; note that upper case letters are used for the variables when written as functions of s and that this does not mean that V is multiplied by s.

2   A constant k which does not vary with time remains a constant. Thus kv, where v is a function of time, becomes $kV(s)$. For example, the voltage $3v$ written as an s function is $3V(s)$.

3   If the initial value of the variable v is zero at time $t = 0$, the first derivative of a function of time dv/dt becomes $sV(s)$ and kdv/dt becomes $ksV(s)$. For example, with no initial values $4dv/dt$ as an s function is $4sV(s)$.

    Note that if there is an initial value v_0 at $t = 0$ then the first derivative of a function of time dv/dt becomes $sV(s) - v_0$, i.e. we subtract any initial value, and kdv/dt becomes $k[sV(s) - v_0]$. For example, if we have $v_0 = 2$ at $t = 0$ then dv/dt becomes $sV(s) - 2$.

4   If the initial value of the variable v and dv/dt is zero at time $t = 0$, the second derivative of a function of time d^2v/dt^2 becomes $s^2V(s)$ and kd^2v/dt^2 becomes $ks^2V(s)$. For example, with no initial values $4d^2v/dt^2$ as an s function is $4s^2V(s)$.

    Note that if there are initial values v_0 and $(dv/dt)_0$ then the second derivative of a function of time d^2v/dt^2 becomes $s^2V(s) - sv_0 - (dv/dt)_0$ and kd^2v/dt^2 becomes $k[s^2V(s) - sv_0 - (dv/dt)_0]$. For example, with initial values of $v_0 = 2$ and $(dv/dt)_0 = 3$ at time $t = 0$, then $4d^2v/dt^2$ as an s function is $4s^2V(s) - 2s - 3$.

5   With an integral of a function of time:

$$\int_0^t v \, dt \text{ becomes } \tfrac{1}{s}V(s)$$

$$\int_0^t kv \, dt \text{ becomes } \tfrac{1}{s}kV(s)$$

Note that, when derivatives are involved, we need to know the initial conditions of a system output prior to the input being applied before we can transform a time function into an s function.

Example

Determine the Laplace transform for the following equation where we have v and v_C as functions of time and no initial values.

$$v = RC\frac{dv_C}{dt} + v_C$$

The Laplace transform is:

$$V(s) = RCsV_C(s) + V_C(s)$$

Thus $V(s)$ is the Laplace transform of the input voltage v and $V_C(s)$ is the Laplace transform of the output voltage v_C. Rearranging gives:

$$\frac{V_C(s)}{V(s)} = \frac{1}{RCs+1}$$

The above equation thus describes the relationship between the input and output of the system when described as s functions.

2.5.1 Transfer function

In section 2.2 we used the term gain to relate the input and output of a system with gain G = output/input. When we are working with inputs and outputs described as functions of s we define the *transfer function* $G(s)$ as [output $Y(s)$/input $X(s)$] when all initial conditions before we apply the input are zero:

$$G(s) = \frac{Y(s)}{X(s)}$$

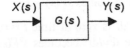

A transfer function can be represented as a block diagram (Figure 2.21) with $X(s)$ the input, $Y(s)$ the output and the transfer function $G(s)$ as the operator in the box that converts the input to the output. The block represents a multiplication for the input. Thus, by using the Laplace transform of inputs and outputs, we can use the transfer function as a simple multiplication factor, like the gain discussed in Section 2.2.

Figure 2.21 *Transfer function as the factor that multiplies the input to give the output*

Example

Determine the transfer function for an electrical system for which we have the relationship (this equation was derived in the example in the preceding section):

$$\frac{V_C(s)}{V(s)} = \frac{1}{RCs+1}$$

The transfer function $G(s)$ is thus:

$$G(s) = \frac{V_C(s)}{V(s)} = \frac{1}{RCs+1}$$

To get the output $V_C(s)$ we multiply the input $V(s)$ by $1/(RCs + 1)$.

Example

Determine the transfer function for the mechanical system, having mass, stiffness and damping, and input F and output y and described by the differential equation (as in Section 2.3.1):

$$F = m\frac{d^2y}{dt^2} + c\frac{dy}{dt} + ky$$

If we now write the equation with the input and output as functions of s, with initial conditions zero:

$$F(s) = ms^2Y(s) + csY(s) + kY(s)$$

Hence the transfer function $G(s)$ of the system is:

$$G(s) = \frac{Y(s)}{F(s)} = \frac{1}{ms^2 + cs + k}$$

2.5.2 Transfer functions of common system elements

By considering the relationships between the inputs to systems and their outputs we can obtain transfer functions for them and hence describe a control system as a series of interconnected blocks, each having its input–output characteristics defined by a transfer function. The following are transfer functions which are typical of commonly encountered system elements:

1 *Gear train*
   For the relationship between the input speed and output speed with a gear train having a gear ratio N:

   transfer function = N

2 *Amplifier*
   For the relationship between the output voltage and the input voltage with G as the constant gain:

   transfer function = G

3 *Potentiometer*
   For the potentiometer acting as a simple potential divider circuit the relationship between the output voltage and the input voltage is the ratio of the resistance across which the output is tapped to the total resistance across which the supply voltage is applied and so is a constant and hence the transfer function is a constant K:

   transfer function = K

4 *Armature-controlled d.c. motor*
   For the relationship between the drive shaft speed and the input voltage to the armature is:

   transfer function = $\dfrac{1}{sL + R}$

where L represents the inductance of the armature circuit and R its resistance.

This was derived by considering armature circuit as effectively inductance in series with resistance and hence:

$$v = L\frac{di}{dt} + Ri$$

and so, with no initial conditions:

$$(V)(s) = sLI(s) + RI(s)$$

and, since the output torque is proportional to the armature current we have a transfer function of the form $1/(sL + R)$.

5   *Valve controlled hydraulic actuator*
The output displacement of the hydraulic cylinder is related to the input displacement of the valve shaft by a transfer function of the form:

$$\text{transfer function} = \frac{K_1}{s(K_2 s + K_3)}$$

where K_1, K_2 and K_3 are constants.

6   *Heating system*
The relationship between the resulting temperature and the input to a heating element is typically of the form:

$$\text{transfer function} = \frac{1}{sC + 1/R}$$

where C is a constant representing the thermal capacity of the system and R a constant representing its thermal resistance.

7   *Tachogenerator*
The relationship between the output voltage and the input rotational speed is likely to be a constant K and so represented by:

$$\text{transfer function} = K$$

8   *Displacement and rotation*
For a system where the input is the rotation of a shaft and the output, as perhaps the result of the rotation of a screw, a displacement, since speed is the rate of displacement we have $v = dy/dt$ and so $V(s) = sY(s)$ and the transfer function is:

$$\text{transfer function} = \frac{1}{s}$$

9   *Height of liquid level in a container*
The height of liquid in a container depends on the rate at which liquid enters the container and the rate at which it is leaving. The

relationship between the input of the rate of liquid entering and the height of liquid in the container is of the form:

$$\text{transfer function} = \frac{1}{sA + \rho g/R}$$

where A is the constant cross-sectional area of the container, ρ the density of the liquid, g the acceleration due to gravity and R the hydraulic resistance offered by the pipe through which the liquid leaves the container.

2.5.3 Transfer functions and systems

Consider a speed control system involving a differential amplifier to amplify the error signal and drive a motor, this then driving a shaft via a gear system. Feedback of the rotation of the shaft is via a tachogenerator.

1   The differential amplifier might be assumed to give an output directly proportional to the error signal input and so be represented by a constant transfer function K, i.e. a gain K which does not change with time.

2   The error signal is an input to the armature circuit of the motor and results in the motor giving an output torque which is proportional to the armature current. The armature circuit can be assumed to be a circuit having inductance L and resistance R and so a transfer function of $1/(sL + R)$.

3   The torque output of the motor is transformed to rotation of the drive shaft by a gear system and we might assume that the rotational speed is proportional to the input torque and so represent the transfer function of the gear system by a constant transfer function N, i.e. the gear ratio.

4   The feedback is via a tachogenerator and we might make the assumption that the output of the generator is directly proportional to its input and so represent it by a constant transfer function H.

The block diagram of the control system might thus be like that in Figure 2.22.

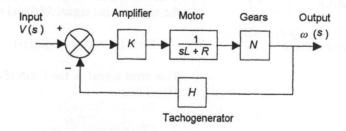

Figure 2.22   *Block diagram for the control system for speed of a shaft with the terms in the boxes being the transfer functions for the elements concerned*

2.6 System transfer functions

Consider the overall transfer functions of systems involving series-connected elements and systems with feedback loops.

2.6.1 Systems in series

Consider a system of two subsystems in series (Figure 2.23). The first subsystem has an input of $X(s)$ and an output of $Y_1(s)$; thus, $G_1(s) = Y_1(s)/X(s)$. The second subsystem has an input of $Y_1(s)$ and an output of $Y(s)$; thus, $G_2(s) = Y(s)/Y_1(s)$. We thus have:

$$Y(s) = G_2(s)Y_1(s) = G_2(s)G_1(s)X(s)$$

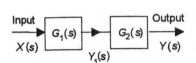

Figure 2.23   *Systems in series*

The overall transfer function $G(s)$ of the system is $Y(s)/X(s)$ and so:

$$G_{overall}(s) = G_1(s)G_2(s)$$

Thus, in general:

The overall transfer function for a system composed of elements in series is the product of the transfer functions of the individual series elements.

Example

Determine the overall transfer function for a system which consists of two elements in series, one having a transfer function of $1/(s + 1)$ and the other $1/(s + 2)$.

The overall transfer function is thus:

$$G_{overall}(s) = \frac{1}{s+1} \times \frac{1}{s+2} = \frac{1}{(s+1)(s+2)}$$

2.6.2 Systems with feedback

For systems with a negative feedback loop we can have the situation shown in Figure 2.24 where the output is fed back via a system with a transfer function $H(s)$ to subtract from the input to the system $G(s)$. The feedback system has an input of $Y(s)$ and thus an output of $H(s)Y(s)$. Thus the feedback signal is $H(s)Y(s)$. The error is the difference between the system input signal $X(s)$ and the feedback signal and is thus:

$$\text{Error }(s) = X(s) - H(s)Y(s)$$

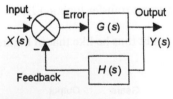

Figure 2.24   *System with negative feedback*

This error signal is the input to the $G(s)$ system and gives an output of $Y(s)$. Thus:

$$G(s) = \frac{Y(s)}{X(s) - H(s)Y(s)}$$

and so:

$$[1 + G(s)H(s)]Y(s) = G(s)X(s)$$

which can be rearranged to give:

$$\text{overall transfer function} = \frac{Y(s)}{X(s)} = \frac{G(s)}{1 + G(s)H(s)}$$

For a system with a negative feedback, the overall transfer function is the forward path transfer function divided by one plus the product of the forward path and feedback path transfer functions.

For a system with positive feedback (Figure 2.25), the feedback signal is $H(s)Y(s)$ and thus the input to the $G(s)$ system is $X(s) + H(s)Y(s)$. Hence:

$$G(s) = \frac{Y(s)}{X(s) + H(s)Y(s)}$$

and so:

$$[1 - G(s)H(s)]Y(s) = G(s)X(s)$$

This can be rearranged to give:

$$\text{overall transfer function} = \frac{Y(s)}{X(s)} = \frac{G(s)}{1 - G(s)H(s)}$$

For a system with a positive feedback, the overall transfer function is the forward path transfer function divided by one minus the product of the forward path and feedback path transfer functions.

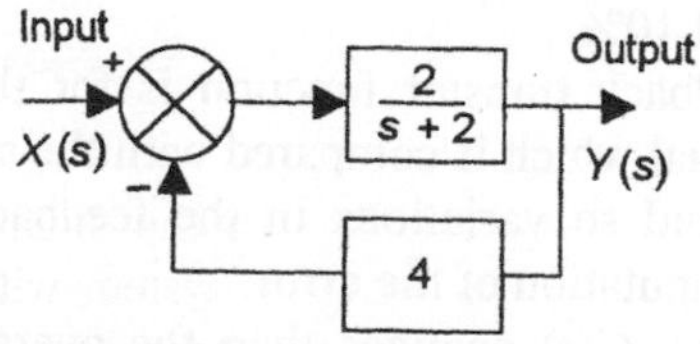

Example

Determine the overall transfer function for a control system (Figure 2.26) which has a negative feedback loop with a transfer function 4 and a forward path transfer function of $2/(s + 2)$.

The overall transfer function of the system is:

$$G_{\text{overall}}(s) = \frac{\dfrac{2}{s+2}}{1 + 4 \times \dfrac{2}{s+2}} = \frac{2}{s + 10}$$

Example

Determine the overall transfer function for a system (Figure 2.27) which has a positive feedback loop with a transfer function 4 and a forward path transfer function of $2/(s + 2)$.

The overall transfer function is:

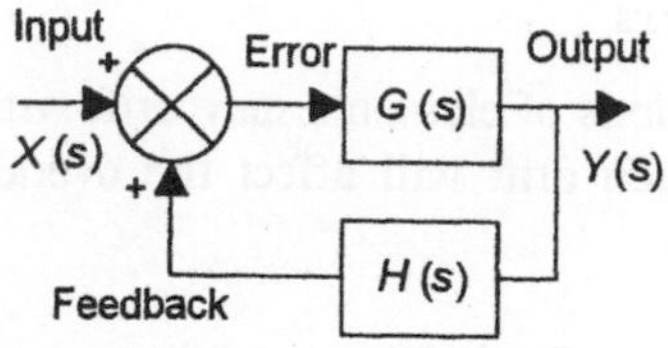

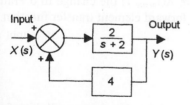

Figure 2.25 *System with positive feedback*

Figure 2.26 *Example*

Figure 2.27 *Example*

$$G_{\text{overall}}(s) = \frac{\frac{2}{s+2}}{1 - 4 \times \frac{2}{s+2}} = \frac{2}{s-6}$$

2.7 Sensitivity

The *sensitivity* of a system is the measure of the amount by which the overall gain of the system is affected by changes in the gain of system elements or particular inputs. In the following, we consider the effects of changing the gain of elements and also the effect of disturbances.

2.7.1 Sensitivity to changes in parameters

With a control system, the transfer functions of elements may drift with time and thus we need to know how such drift will affect the overall performance of the system.

For a closed-loop system with negative feedback (Figure 2.28):

$$\text{overall transfer function} = \frac{G(s)}{1 + G(s)H(s)}$$

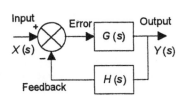

Figure 2.28   *System with negative feedback*

If $G(s)H(s)$ is large then the above equation reduces to:

$$\text{overall transfer function} \simeq \frac{G(s)}{G(s)H(s)} \simeq \frac{1}{H(s)}$$

Thus, in such a situation, the system is relatively insensitive to variations in the forward path transfer function but is sensitive to variations in the feedback path transfer function. For example, a change in the feedback path transfer function of, say, 10%, i.e. from $H(s)$ to $1.1H(s)$, will result in a change in the overall transfer function from $1/H(s)$ to $1/1.1H(s)$ or about $0.9/H(s)$ and so a change of about 10%.

This sensitivity is because the feedback transfer function is for the measurement system supplying the signal which is compared with the set value signal to determine the error and so variations in the feedback transfer function directly affect the computation of the error.

If the forward path transfer function $G(s)$ changes then the overall transfer function $G_{\text{overlll}}(s)$ will change. We can define the sensitivity of the system to changes in the transfer function of the forward element as the fractional change in the overall system transfer function $G_{\text{overall}}(s)$ divided by the fractional change in the forward element transfer function $G(s)$, i.e. $(\Delta G_{\text{overall}}/G_{\text{overall}})/(\Delta G/G)$ where $\Delta G_{\text{overall}}$ is the change in overall gain producing a change of ΔG in the forward element transfer function. Thus, the sensitivity can be written as:

$$\text{sensitivity} = \frac{\Delta G_{\text{overall}}(s)}{\Delta G(s)} \frac{G(s)}{G_{\text{overall}}(s)}$$

If we differentiate the equation given above for the overall transfer function we obtain:

$$\frac{dG_{overall}(s)}{dG(s)} = \frac{1}{\left[1+G(s)H(s)\right]^2}$$

and since $G_{overall}(s)/G(s) = 1/[1 + G(s)H(s)]$, the sensitivity is:

$$\text{sensitivity} = \frac{1}{1+G(s)H(s)}$$

Thus the bigger the value of $G(s)H(s)$ the lower the sensitivity of the system to changes in the forward path transfer function. The feedback amplifier discussed in Section 2.2.3 is an illustration of this, the forward path transfer function for the op amp being very large and so giving a system with low sensitivity to changes in the op amp gain and hence a stable system which can have its gain determined by purely changing the feedback loop gain, i.e. the resistors in a potential divider.

Example

A closed-loop control system with negative feedback has a feedback transfer function of 0.1 and a forward path transfer function of (a) 50, (b) 5. What will be the effect of a change in the forward path transfer function of an increase by 10%?

(a) We have, before the change:

$$\text{overall transfer function} = \frac{G(s)}{1+G(s)H(s)} = \frac{50}{1+50\times0.1} = 8.3$$

After the change we have:

$$\text{overall transfer function} = \frac{G(s)}{1+G(s)H(s)} = \frac{55}{1+55\times0.1} = 8.5$$

The change is thus about 2%.
(b) We have, before the change:

$$\text{overall transfer function} = \frac{G(s)}{1+G(s)H(s)} = \frac{5}{1+5\times0.1} = 3.3$$

After the change we have:

$$\text{overall transfer function} = \frac{G(s)}{1+G(s)H(s)} = \frac{5.5}{1+5.5\times0.1} = 3.5$$

The change is thus about 6%.
Thus the sensitivity of the system to changes in the forward path transfer function is reduced as the gain of the forward path is increased.

2.7.2 Sensitivity to disturbances

An important effect of having feedback in a system is the reduction of the effects of disturbance signals on the system. A disturbance signal is an unwanted signal which affects the output signal of a system, e.g. noise in electronic amplifiers or a door being opened in a room with temperature controlled by a central heating system.

Consider the effect of external disturbances on the overall gain of a system. Firstly we consider the effect on a open-loop system and then on a closed-loop system.

Consider the two-element open-loop system shown in Figure 2.29 when there is a disturbance which gives an input between the two elements. For an input $X(s)$ to the system, the first element gives an output of $G_1(s)X(s)$. To this is added the disturbance $D(s)$ to give an input of $G_1(s)X(s) + D(s)$. The overall system output will then be

$$Y(s) = G_2(s)[G_1(s)X(s) + D(s)] = G_1(s)G_2(s)X(s) + G_2(s)D(s)$$

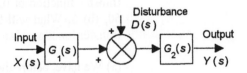

Figure 2.29   *Disturbance with an open-loop system*

The *signal-to-noise ratio* is the ratio of the output due to the signal to that produced by the noise and is thus, for the open-loop system, given by $G_1(s)G_2(s)X(s)/G_2(s)D(s) = G_1(s)X(s)/D(s)$. Thus increasing the gain of the element prior to the disturbance increases the signal-to-noise ratio.

For the system with feedback (Figure 2.30), the input to the first forward element $G_1(s)$ is $X(s) - H(s)Y(s)$ and so its output is $G_1(s)[X(s) - H(s)Y(s)]$. The input to $G_2(s)$ is $G_1(s)[X(s) - H(s)Y(s)] + D(s)$ and so its output is $X(s) = G_2(s)\{G_1(s)[X(s) - H(s)Y(s)] + D(s)\}$. Rearranged this gives:

$$Y(s) = \frac{G_1(s)G_2(s)}{1 + G_1(s)G_2(s)H(s)}X(s) + \frac{G_2(s)}{1 + G_1(s)G_2(s)H(s)}D(s)$$

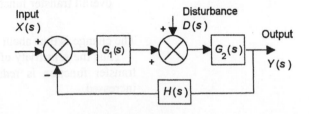

Figure 2.30   *Disturbance with closed-loop system*

Comparing this with the equation for the open-loop system of $Y(s) = G_2(s)[G_1(s)X(s) + D(s)] = G_1(s)G_2(s)X(s) + G_2(s)D(s)$ indicates that the effect of the disturbance on the output of the system has been reduced by a factor of $[1 + G_1(s)G_2(s)H(s)]$. This factor is thus a measure of how much the effects of a disturbance are reduced by feedback.

The signal-to-noise ratio is the ratio of the output due to the signal to that produced by the noise and is thus $G_1(s)X(s)/D(s)$ and is the same as when there is no feedback. Thus, as with the open-loop system, the effect of such a disturbance is reduced as the gain of the $G_1(s)$ element is increased.

2.8 Block manipulation

The following are some of the ways we can reorganise the blocks in a block diagram of a system in order to produce simplification and still give the same overall transfer function for the system. To simplify the diagrams, the (s) has been omitted; it should, however, be assumed for all dynamic situations.

2.8.1 Blocks in series

As indicated in Section 2.5.1, Figure 2.31 shows the basic rule for simplifying blocks in series.

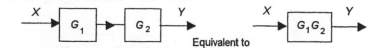

Figure 2.31 *Blocks in series*

2.8.2 Moving take-off points

As a means of simplifying block diagrams it is often necessary to move takeoff points. The following figures (Figures 2.32 and 2.33) give the basic rules for such movements.

Figure 2.32 *Moving a takeoff point to beyond a block*

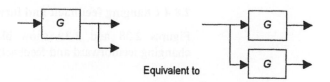

Figure 2.33 *Moving a takeoff point to ahead of a block*

2.8.3 Moving a summing point

As a means of simplifying block diagrams it is often necessary to move summing points. The following figures (Figures 2.34–2.37) give the basic rules for such movements.

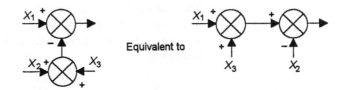

Figure 2.34   *Rearrangement of summing points*

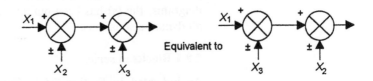

Figure 2.35   *Interchange of summing points*

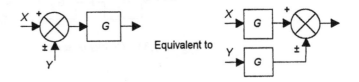

Figure 2.36   *Moving a summing point ahead of a block*

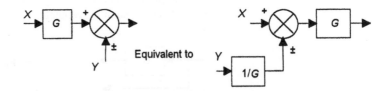

Figure 2.37   *Moving a summing point beyond a block*

2.8.4 Changing feedback and forward paths

Figures 2.38 and 2.39 show block simplification techniques when changing feedforward and feedback paths.

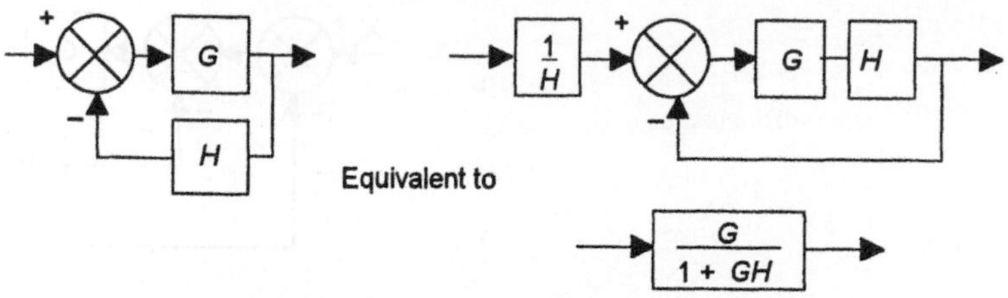

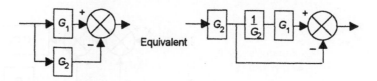

Figure 2.38   *Removing a block from a feedback path*

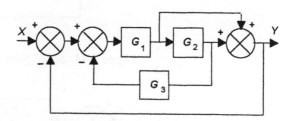

Figure 2.39   *Removing a block from a forward path*

Example

Use block simplification techniques to simplify the system shown in Figure 2.40.

Figures 2.41–2.46 show the various stages in the simplification.

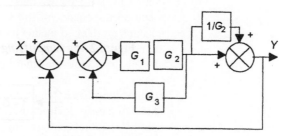

Figure 2.40   *The circuit to be simplified*

Figure 2.41   *Moving a takeoff point*

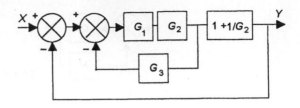

Figure 2.42   *Eliminating a feedforward loop*

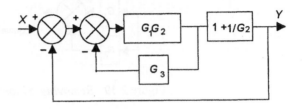

Figure 2.43   *Simplifying series elements*

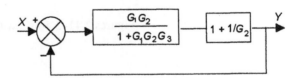

Figure 2.44   *Simplifying a feedback element*

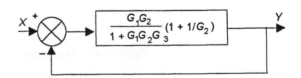

Figure 2.45   *Simplifying series elements*

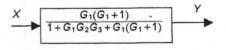

Figure 2.46   *Simplifying negative feedback*

2.9 Multiple inputs   When there is more than one input to a system, the *superpositio.*
principle can be used. This is that *the response to several input*

simultaneously applied is the sum of the individual responses to each input when applied separately. Thus, the procedure to be adopted for a multi-input-single output (MISO) system is:

1 Set all but one of the inputs to zero.

2 Determine the output signal due to this one non-zero input.

3 Repeat the above steps for each input in turn.

4 The total output of the system is the algebraic sum of the outputs due to each of the inputs.

Example

Determine the output $Y(s)$ of the system shown in Figure 2.47 when there is an input $X(s)$ to the system as a whole and a disturbance signal $D(s)$ at the point indicated.

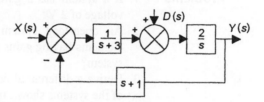

Figure 2.47 *System with a disturbance input*

If we set $D(s)$ to zero we have the system shown in Figure 2.48 and the output is given by:

$$\frac{Y(s)}{X(s)} = \frac{2}{s(s+3) + 2(s+1)}$$

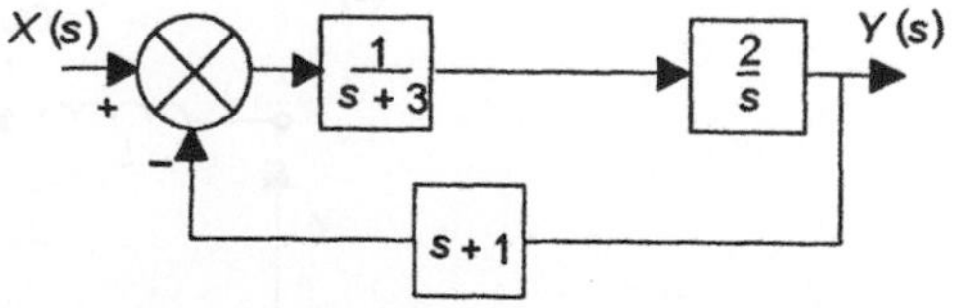

Figure 2.48 *System with disturbance put equal to zero*

If we now set $X(s)$ to zero we have the system shown in Figure 2.49. This is a system with a forward path transfer function of $2/s$ and a positive feedback of $(1/s + 3)[-(s + 1)]$. This gives an output of:

$$\frac{Y(s)}{D(s)} = \frac{2(s+3)}{s(s+3) + 2(s+1)}$$

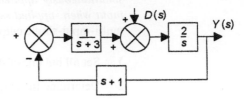

Figure 2.49   *System with input equal to zero*

The total input is the sum of the outputs due to each of the inputs and so:

$$Y(s) = \frac{2}{s(s+3)+2(s+1)}X(s) + \frac{2(s+3)}{s(s+3)+2(s+1)}D(s)$$

Problems

1 If a system has a gain of 5, what will be the output for an input voltage of 2 V?

2 An open-loop system consists of three elements in series, the elements having gains of 2, 5 and 10. What is the overall gain of the system?

3 Derive a differential equation relating the input and output for each of the systems shown in Figure 2.50.

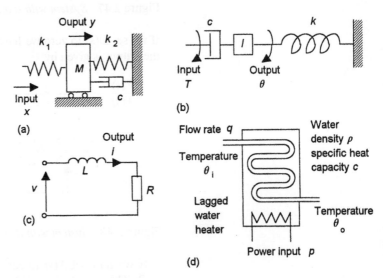

Figure 2.50 *Problem 3*

4 An open loop system consists of three elements in series, the elements having transfer functions of 5, $1/s$ and $1/(s + 1)$. What is the overall transfer function of the system?

5 What is the overall gain of a closed-loop negative feedback system having a forward path gain of 2 and a feedback path gain of 0.1?

6 What is the overall transfer function of a closed-loop negative feedback system having a forward path transfer function of $2/(s + 1)$ and a feedback path transfer function of 0.1?

7 Figure 2.51 shows an electrical circuit and its block diagram representation. What is the overall transfer function of the system?

8 Use block simplification to arrive at the overall transfer function of the systems shown in Figure 2.52.

9 What is the overall transfer function for the systems shown in Figure 2.52?

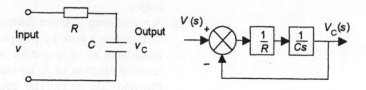

Figure 2.51 *Problem 7*

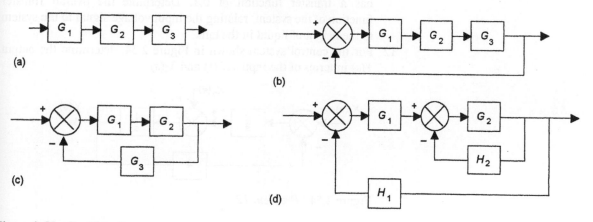

(a)

(b)

(c)

(d)

Figure 2.52 *Problem 8*

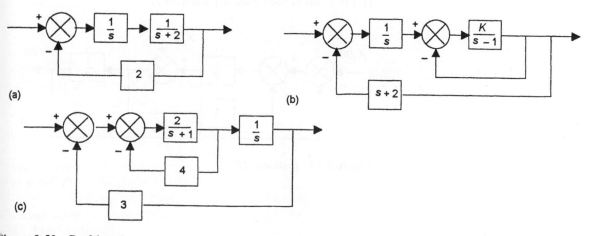

(a)

(b)

(c)

Figure 2.53 *Problem 9*

10   A closed-loop negative feedback system to be used for controlling
the position of a load has a differential amplifier with transfer
function K_1 operating a motor with transfer function $1/(sL + R)$. The
output of the motor operates a gear system with gear ratio N and
this, in turn, operates a screw with transfer function $1/s$ to give the
resulting displacement. The position sensor is a potentiometer and
this gives a feedback voltage related to the position of the load by the
transfer function K_2. Derive the transfer function for the system as a
whole, relating the input voltage to the system to the displacement
output.

11   A closed-loop negative feedback system for the control of the height
of liquid in a tank by pumping liquid from a reservoir tank can be
considered to be a system with a differential amplifier having a
transfer function of 5, its output operating a pump with a transfer
function $5/(s + 1)$. The coupled system of tanks has a transfer
function, relating height in the tank to the output from the pump, of
$3/(s + 1)(s + 2)$. The feedback sensor of the height level in the tank
has a transfer function of 0.1. Determine the overall transfer
function of the system, relating the input voltage signal to the system
to the height of liquid in the tank.

12   For the control system shown in Figure 2.54, determine the output
$Y(s)$ in terms of the inputs $X_1(s)$ and $X_2(s)$.

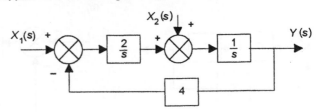

Figure 2.54   *Problem 12*

13   For the control system shown in Figure 2.55, determine the output
$Y(s)$ in terms of the inputs $X_1(s)$ and $X_2(s)$.

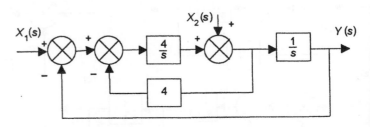

Figure 2.55   *Problem 13*

3 System response

3.1 Introduction

How will the output from a system change when there is an input to it? For example, if we consider a mercury-in-glass thermometer as a system with an input of temperature and an output of the level of the mercury in the glass capillary, how will the level change when the thermometer is suddenly immersed in hot water, i.e. given a step input? As a matter of experience we know that the level will increases as a result of the input and take a certain amount of time before it reaches its steady state value. If we consider a control system employing a motor and feedback to move a work piece in an automatic machining operation, how will the output, i.e. the displacement of the workpiece, vary with time when the input is gradually increased with time with the aim of gradually increasing the displacement of the workpiece? This chapter is concerned with a method we can use to answer this question and, in general, determine how the output of systems changes when there is a change in input.

In Chapter 2 the method of describing a system by means of a transfer function was introduced; the transfer function is the output divided by the input when both are written as s functions. In this chapter we consider how we can use transfer functions to determine how the output of a system will change with time for particular inputs.

3.2 Inputs

Inputs to systems commonly take a number of standard forms (Figure 3.1). With the *step input* we have the input suddenly being switched to a constant value at some particular time. With the *impulse input* we have the input existing for just a very brief time before dropping back to zero. With the *ramp input*, we have the input starting at some time and then increasing at a constant rate. We can write such inputs as s functions in the following way:

1   A *unit step input* which starts at a time $t = 0$ and rises to the constant value 1 has a Laplace transform of the input as an s function multiplied by $1/s$.

2   A *unit impulse input* which starts at a time $t = 0$ and rises to the value 1 has a Laplace transform of the input as an s function multiplied by 1.

3   A *unit ramp input* which starts at time $t = 0$ and rises by 1 each second has a Laplace transform of the input as an s function multiplied by $1/s^2$.

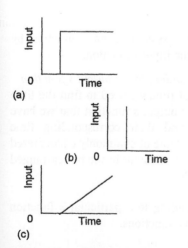

Figure 3.1   *Forms of input:*
(a) step, (b) impulse, (c) ramp

In general, if a function of time is multiplied by some constant, then the Laplace transform of that function is multiplied by the same constant. Thus, if we have a step input of size 5 then the Laplace transform is 5 times the transform of a unit step and so the input as an s function is $5/s$.

Example

An electrical system has an input of a voltage of 2 V which is suddenly applied by a switch being closed. What is the input as an s function?

Assume the input occurs at time $t = 0$. The input is a step voltage of size 2 V. An input of a unit step voltage as an s function is $(1/s)$ and thus for a 2 V step is $(2/s)$ V.

Example

A controlled speed motor has a voltage input which is increased at the rate of 3 V per second. What is the input as an s function?

Assume the input starts at time $t = 0$. The input is a ramp voltage of 3 V/s. An input of a unit ramp voltage as an s function is $(1/s^2)$ and thus for a 3 V/s ramp is $(3/s^2)$.

3.3 Determining outputs

The procedure we can use to determine how the output of a system will change with time when there is some input to the system is:

1   *Determine the output as an s function*
    In terms of the transfer function $G(s)$ we have:

    $$\text{Output } (s) = G(s) \times \text{Input } (s)$$

    We can thus obtain the output of a system as an s function by multiplying its transfer function by the input s function.

2   *Determine the time function corresponding to the output s function*
    To obtain the output as a function of time we need to find the time function that will give the particular output s function that we have obtained. Tables of s functions and their corresponding time functions can be used; Table 3.1 is a table of commonly encountered functions. Often, however, the s function output has to be rearranged to put it into a form given in the table.

In obtaining the time function corresponding to a particular s function we can utilise the following properties of s functions:

1   If a Laplace transform is multiplied by some constant then the corresponding time function is multiplied by the same constant. For example, if we have $3/(s + 1)$ then the corresponding time function is that of 3 times the time function for $1/(s + 1)$.

Table 3.1 *Laplace functions and their corresponding time functions*

	Time function $f(t)$	Laplace transform $F(s)$
1	A unit impulse	1
2	A unit step	$\dfrac{1}{s}$
3	t, a unit ramp	$\dfrac{1}{s^2}$
4	e^{-at}, exponential decay	$\dfrac{1}{s+a}$
5	$1 - e^{-at}$, exponential growth	$\dfrac{a}{s(s+a)}$
6	te^{-at}	$\dfrac{1}{(s+a)^2}$
7	$t - \dfrac{1-e^{-at}}{a}$	$\dfrac{a}{s^2(s+a)}$
8	$e^{-at} - e^{-bt}$	$\dfrac{b-a}{(s+a)(s+b)}$
9	$(1 - at)e^{-at}$	$\dfrac{s}{(s+a)^2}$
10	$1 - \dfrac{b}{b-a}e^{-at} + \dfrac{a}{b-a}e^{-bt}$	$\dfrac{ab}{s(s+a)(s+b)}$
11	$\dfrac{e^{-at}}{(b-a)(c-a)} + \dfrac{e^{-bt}}{(c-a)(a-b)} + \dfrac{e^{-ct}}{(a-c)(b-c)}$	$\dfrac{1}{(s+a)(s+b)(s+c)}$
12	$\sin \omega t$, a sine wave	$\dfrac{\omega}{s^2+\omega^2}$
13	$\cos \omega t$, a cosine wave	$\dfrac{s}{s^2+\omega^2}$
14	$e^{-at}\sin \omega t$, a damped sine wave	$\dfrac{\omega}{(s+a)^2+\omega^2}$
15	$e^{-at}\cos \omega t$, a damped cosine wave	$\dfrac{s+a}{(s+a)^2+\omega^2}$
16	$\dfrac{\omega}{\sqrt{1-\zeta^2}}e^{-\zeta\omega t}\sin \omega\sqrt{1-\zeta^2}\,t$	$\dfrac{\omega^2}{s^2+2\zeta\omega s+\omega^2}$
17	$1 - \dfrac{1}{\sqrt{1-\zeta^2}}e^{-\zeta\omega t}\sin\left(\omega\sqrt{1-\zeta^2}\,t + \phi\right),\ \cos \phi = \zeta$	$\dfrac{\omega^2}{s(s^2+2\zeta\omega s+\omega^2)}$

2   If we have two separate s terms then the corresponding time function is the sum of the time functions corresponding to each of the separate s terms. For example, if we have $[1/(s + 2)] + [1/(s + 1)]$ then the time function is the time function for $[1/(s + 2)]$ plus the time function for $[1/(s + 1)]$.

Example

A system gives an output of $1/(s + 5)$. What is the output as a function of time?

The output is of the form given in Table 3.1 as item 4 with $a = 5$. Hence the time function is e^{-5t} and thus describes an output which decays exponentially with time.

Example

A system gives an output of $10/[s(s + 5)]$. What is the output as a function of time?

The nearest form we have in Table 3.1 to the output is item 5 as $2 \times a/[s(s + a)]$ with $a = 5$. Thus the output, as a function of time, is $2(1 - e^{-5t})$.

Example

A system has a transfer function of $1/(s + 2)$. What will be its output as a function of time when it is subject to a step input of 1 V?

The step input has a Laplace transform of $(1/s)$. Thus:

$$\text{Output } (s) = G(s) \times \text{Input } (s)$$

$$= \frac{1}{s+2} \times \frac{1}{s} = \frac{1}{s(s+2)}$$

The nearest form we have in Table 3.1 to the output is item 5 as $\frac{1}{2} \times 2/[s(s + 2)]$. Thus the output, as a function of time, is $\frac{1}{2}(1 - e^{-5t})$ V.

Example

A system has a transfer function of $4/(s + 2)$. What will be its output as a function of time when subject to a ramp input of 2 V/s?

The ramp input has a Laplace transform of $(2/s^2)$. Thus:

$$\text{Output } (s) = G(s) \times \text{Input } (s)$$

$$= \frac{4}{s+2} \times \frac{2}{s^2} = \frac{8}{s^2(s+2)}$$

The nearest form we have in Table 3.1 to the output is item 7 when written as $4 \times 2/[s^2(s + 2)]$. Thus the output, as a function of time, is $4[t - (1 - e^{-2t})/2] = 4t - 2(1 - e^{-2t})$ V.

3.3.1 Partial fractions

A technique that is often required to put an s function in terms which identify with forms, so enabling the corresponding time function to be obtained in Table 3.1 is *partial fractions*. The term partial fractions is used for the process of converting an expression involving a complex fraction into a number of simpler fraction terms. This technique can be used provided the highest power of *s* in the numerator of the expression is less than that in the denominator. When the highest power in the numerator is equal to or higher than that of the denominator, the denominator must be divided into the numerator until the result is the sum of terms with the remainder fractional term having a numerator with a lower power than the denominator.

There are basically three types of partial fractions:

1   The numerator is some function of *s* and the denominator contains factors which are only of the form $(s + a)$, $(s + b)$, $(s + c)$, etc. and so is of the form:

$$\frac{F(s)}{(s+a)(s+b)(s+c)}$$

and has the partial fractions of

$$\frac{A}{(s+a)} + \frac{B}{(s+b)} + \frac{C}{(s+c)}$$

There is a partial fraction term for each bracketed term in the denominator. Thus, if we have $1/(s + a)(s + b)$ there will be two partial fraction terms.

2   There are repeated $(s + a)$ factors in the denominator and the expression is of the form:

$$\frac{F(s)}{(s+a)^n}$$

and has the partial fractions of:

$$\frac{A}{(s+a)^1} + \frac{B}{(s+a)^2} + \frac{C}{(s+a)^3} + \cdots + \frac{N}{(s+a)^n}$$

A multiple root expression has thus a partial fraction term for each power of the factor. Thus, if we have $1/(s + a)^2$ there will be two partial fraction terms; if we have $1/(s + a)^3$ there will be three partial fraction terms.

3   The denominator contains quadratic factors and the quadratic does not factorise, being of the form:

$$\frac{F(s)}{(as^2 + bs + c)(s + d)}$$

and has the partial fractions of:

$$\frac{As+B}{as^2 + bs + c} + \frac{C}{s + d}$$

The values of the constants A, B, C, etc. can be found by either making use of the fact that the equality between the expression and the partial fractions must be true for all values of s and so considering particular values of s that make calculations easy or that the coefficients of s^n in the expression must equal those of s^n in the partial fraction expansion.

Example

Determine the partial fractions of:

$$\frac{s+4}{(s+1)(s+2)}$$

The partial fractions are of the form:

$$\frac{A}{s+1} + \frac{B}{s+2}$$

Then, for the partial fraction expression to equal the original fraction, we must have:

$$\frac{s+4}{(s+1)(s+2)} = \frac{A(s+2)+B(s+1)}{(s+1)(s+2)}$$

and consequently:

$$s+4 = A(s+2) + B(s+1)$$

This must be true for all values of s. The procedure is then to pick values of s that will enable some of the terms involving constants to become zero and so enable other constants to be determined. Thus if we let $s = -2$ then we have

$$(-2) + 4 = A(-2+2) + B(-2+1)$$

and so $B = -2$. If we now let $s = -1$ then

$$(-1) + 4 = A(-1+2) + B(-1+1)$$

and so $A = 3$. Thus

$$\frac{s+4}{(s+1)(s+2)} = \frac{3}{s+1} - \frac{2}{s+2}$$

Example

Determine the partial fractions of:

$$\frac{3s+1}{(s+2)^3}$$

This has partial fractions of:

$$\frac{A}{(s+2)} + \frac{B}{(s+2)^2} + \frac{C}{(s+2)^3}$$

Then, for the partial fraction expression to equal the original fraction, we must have:

$$\frac{3s+1}{(s+2)^3} = \frac{A}{(s+2)} + \frac{B}{(s+2)^2} + \frac{C}{(s+2)^3}$$

and so consequently have:

$$3s + 1 = A(s+2)^2 + B(s+2) + C$$

$$= A(s^2 + 2s + 1) + B(s+2) + C$$

Equating s^2 terms gives $0 = A$. Equating s terms gives $3 = 2A + B$. and so $B = 3$. Equating the numeric terms gives $1 = A + 2B + C$ and so $C = -5$. Thus:

$$\frac{3s+1}{(s+2)^3} = \frac{3}{(s+2)^2} - \frac{5}{(s+2)^3}$$

Example

Determine the partial fractions of:

$$\frac{2s+1}{(s^2+s+1)(s+2)}$$

This will have partial fractions of:

$$\frac{As+B}{s^2+s+1} + \frac{C}{s+2}$$

Thus we must have:

$$\frac{2s+1}{(s^2+s+1)(s+2)} = \frac{As+B}{s^2+s+1} + \frac{C}{s+2}$$

and so:

$$2s + 1 = (As + B)(s+2) + C(s^2 + s + 1)$$

With $s = -2$ then $-3 = 3C$ and so $C = -1$. Equating s^2 terms gives $0 = A + C$ and so $A = 1$. Equating s terms gives $2 = 2A + B + C$ and so $B = 1$. As a check, equating numeric terms gives $1 = 2B + C$. Thus:

$$\frac{2s+1}{(s^2+s+1)(s+2)} = \frac{s+1}{s^2+s+1} - \frac{1}{s+2}$$

3.4 First order systems

A first order system has a differential equation of the form (see Section 2.3.3):

$$\tau \frac{dy}{dt} + y = kx$$

As a function of s this can be written as:

$$\tau Y(s) + Y(s) = kX(s)$$

and so a transfer function of the form:

$$G(s) = \frac{Y(s)}{X(s)} = \frac{k}{\tau s + 1}$$

where k is the *gain* of the system when there are steady-state conditions and τ is the *time constant* of the system.

When a first-order system is subject to a unit impulse input then $X(s) = 1$ and the output transform $Y(s)$ is:

$$Y(s) = G(s)X(s) = \frac{k}{\tau s + 1} \times 1 = k\frac{(1/\tau)}{s + 1/\tau}$$

Hence, since we have the transform in the form $1/(s + a)$, using item 4 in Table 3.1 gives:

$$x = k(1/\tau)\, e^{-t/\tau}$$

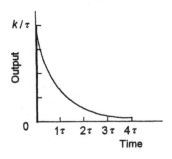

k/τ

Output

0   1τ   2τ   3τ   4τ
              Time

Figure 3.2   *Output with a unit impulse input to a first order system*

Figure 3.2 shows how the output varies with time; it is an exponential decay. The output rises to its maximum value at time $t = 0$ and then after 1τ it drops to 0.37 of the initial value, after 2τ it is 0.14 of the initial value and after 3τ it is 0.05 of the initial value. Thus by about a time equal to four times the time constant the output is effectively zero. The exponential term tends to a zero value as the time t tends to an infinite value.

When a first-order system is subject to a unit step input then $X(s) = 1/s$ and the output transform $Y(s)$ is:

$$X(s) = G(s)Y(s) = \frac{k}{s(\tau s + 1)} = k\frac{(1/\tau)}{s(s + 1/\tau)}$$

Hence, since we have the transform in the form $a/s(s + a)$, using item 5 in Table 3.1 gives:

$$x = k(1 - e^{-t/\tau})$$

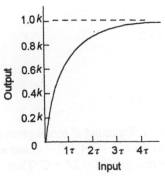

1.0k
0.8k
0.6k
Output
0.4k
0.2k

0   1τ   2τ   3τ   4τ
              Input

Figure 3.3   *Behaviour of a first order system when subject to a unit step input*

Figure 3.3 shows how the output varies with time. Initially, at time $t = 0$, the output is zero. It then rises to 0.63 of the steady state value after 1τ, then 0.86 of the steady state value after 2τ and 0.95 of the steady state value after 3τ. After 4τ the output is effectively at the steady state value of k, the exponential term in the above equation becoming zero as time t tends to infinity.

Example

A circuit has a resistance R in series with a capacitance C. The differential equation relating the input v and output v_C, i.e. the voltage across the capacitor, is:

$$v = RC \frac{dv_C}{dt} + v_C$$

Determine the output of the system when there is a 3 V impulse input.

As a function of s the differential equation becomes:

$$V(s) = RCsV_C(s) + V_C(s)$$

Hence the transfer function is

$$G(s) = \frac{V_C(s)}{V(s)} = \frac{1}{RCs + 1}$$

The output when there is 2 V impulse input is:

$$V_C(s) = G(s)V(s) = \frac{1}{RCs + 1} \times 3 = \frac{3/RC}{s + 1/RC}$$

Hence, since we have the transform in the form $1/(s + a)$, using item 4 in Table 3.1 gives:

$$x = (2/RC)\, e^{-t/RC}$$

Example

A thermocouple which has a transfer function linking its voltage output V and temperature input of:

$$G(s) = \frac{30 \times 10^{-6}}{10s + 1} \ \text{V/°C}$$

Determine the response of the system when it is suddenly immersed in a water bath at 100°C.

The output as an s function is:

$$V(s) = G(s) \times \text{input}\ (s)$$

The sudden immersion of the thermometer gives a step input of size 100°C and so the input as an s function is $100/s$. Thus:

$$V(s) = \frac{30 \times 10^{-6}}{10s+1} \times \frac{100}{s} = \frac{30 \times 10^{-4}}{10s(s+0.1)} = 30 \times 10^{-4} \frac{0.1}{s(s+0.1)}$$

The fraction element is of the form $a/s(s + a)$ and so the output as a function of time is:

$$V = 30 \times 10^{-4} \, (1 - e^{-0.1t}) \text{ V}$$

3.5 Second order systems

The differential equation for a second-order system is of the form:

$$\frac{d^2y}{dt^2} + 2\zeta\omega_n\frac{dy}{dt} + \omega_n^2 y = k\omega_n^2 x$$

where ω_n is the natural angular frequency with which the system oscillates and ζ is the damping ratio. Hence we have:

$$s^2 Y(s) + 2\zeta\omega_n sY(s) + \omega_n^2 Y(s) = k\omega_n^2 X(s)$$

and so a transfer function of:

$$G(s) = \frac{Y(s)}{X(s)} = \frac{k\omega_n^2}{s^2 + 2\zeta\omega_n s + \omega_n^2}$$

When a second-order system is subject to a unit step input, i.e. $X(s) = 1/s$, then the output transform is:

$$Y(s) = G(s)X(s) = \frac{k\omega_n^2}{s(s^2 + 2\zeta\omega_n s + \omega_n)}$$

There are three different forms of answer to this equation for the way the output varies with time; these depending on the value of the damping constant and whether it gives an overdamped, critically damped or underdamped system (see Figure 2.13). We can determine the condition for these three forms of output by putting the equation in the form:

$$Y(s) = \frac{k\omega_n^2}{s(s+p_1)(s+p_2)}$$

where p_1 and p_2 are the roots of the quadratic term:

$$s^2 + 2\zeta\omega_n s + \omega_n^2 = 0$$

Hence, if we use the equation to determine the roots of a quadratic equation, we obtain:

$$p = \frac{-2\zeta\omega_n \pm \sqrt{4\zeta^2\omega_n^2 - 4\omega_n^2}}{2}$$

and so the two roots are given by:

$$p_1 = -\zeta\omega_n + \omega_n\sqrt{\zeta^2 - 1} \quad \text{and} \quad p_2 = -\zeta\omega_n - \omega_n\sqrt{\zeta^2 - 1}$$

The important issue in determining the form of the roots is the value of the square root term.

1   $\zeta > 1$

With the damping factor ζ greater than 1 the square root term is real and will factorise. To find the inverse transform we can either use partial fractions to break the expression down into a number of simple fractions or use item 10 in Table 3.1. The output is thus:

$$y = \frac{k\omega_n^2}{p_1 p_2}\left[1 - \frac{p_2}{p_2 - p_1}\,e^{-p_1 t} + \frac{p_1}{p_2 - p_1}\,e^{-p_2 t}\right]$$

This describes an output which does not oscillate but dies away with time and thus the system is *overdamped*. As the time t tends to infinity then the exponential terms tend to zero and the output becomes the steady value of $k\omega_n^2/(p_1 p_2)$. Since $p_1 p_2 = \omega_n^2$, the steady value is k.

2   $\zeta = 1$

With $\zeta = 1$ the square root term is zero and so $p_1 = p_2 = -\omega_n$; both roots are real and both the same. The output equation then becomes:

$$Y(s) = \frac{k\omega_n^2}{s(s + \omega_n)^2}$$

This equation can be expanded by means of partial fractions to give:

$$Y(s) = k\left[\frac{1}{s} - \frac{1}{s + \omega_n} - \frac{\omega_n}{(s + \omega_n)^2}\right]$$

Hence:

$$y = k[1 - e^{-\omega_n t} - \omega_n t\, e^{-\omega_n t}]$$

This is the critically damped condition and describes an output which does not oscillate but dies away with time. As the time t tends to infinity then the exponential terms tend to zero and the output tends to the steady state value of k.

3   $\zeta < 1$

With $\zeta < 1$ the square root term does not have a real value. Using item 17 in Table 3.1 then gives:

$$y = k\left[1 - \frac{e^{-\zeta\omega_n t}}{\sqrt{1 - \zeta^2}}\,\sin\left(\omega_n\sqrt{(1 - \zeta^2)}\,t + \phi\right)\right]$$

where $\cos\phi = \zeta$. This is an under-damped oscillation. The angular frequency of the damped oscillation is:

$$\omega = \omega_n \sqrt{1 - \zeta^2}$$

Only when the damping is very small does the angular frequency of the oscillation become nearly the natural angular frequency ω_n. As the time t tends to infinity then the exponential term tends to zero and so the output tends to the value k.

Example

What will be the state of damping of a system having the following transfer function and subject to a unit step input?

$$G(s) = \frac{1}{s^2 + 8s + 16}$$

The output $Y(s)$ from such a system is given by:

$$Y(s) = G(s)X(s)$$

For a unit step input $X(s) = 1/s$ and so the output is given by:

$$Y(s) = \frac{1}{s(s^2 + 8s + 16)} = \frac{1}{s(s+4)(s+4)}$$

The roots of $s^2 + 8s + 16$ are $p_1 = p_2 = -4$. Both the roots are real and the same and so the system is critically damped.

Example

A system has an output y related to the input x by the differential equation:

$$\frac{d^2y}{dt^2} + 5\frac{dy}{dt} + 6y = x$$

What will be the output from the system when it is subject to a unit step input? Initially both the output and input are zero.

We can write the Laplace transform of the equation as:

$$s^2Y(s) + 5sY(s) + 6Y(s) = X(s)$$

The transfer function is thus:

$$G(s) = \frac{Y(s)}{X(s)} = \frac{1}{s^2 + 5s + 6}$$

For a unit step input the output is given by:

$$Y(s) = \frac{1}{s(s^2 + 5s + 6)} = \frac{1}{s(s+3)(s+2)}$$

Because the quadratic term has two real roots, the system is overdamped. We can directly use one of the standard forms given in Table 3.1 or partial fractions to first simplify the expression before using Table 3.1. Using partial fractions:

$$\frac{1}{s(s+3)(s+2)} = \frac{A}{s} + \frac{B}{s+3} + \frac{C}{s+2}$$

Thus, we have $1 = A(s + 3)(s + 2) + Bs(s + 2) + Cs(s + 3)$. When $s = 0$ then $1 = 6A$ and so $A = 1/6$. When $s = -3$ then $1 = 3B$ and so $B = 1/3$. When $s = -2$ then $1 = -2C$ and so $C = -1/2$. Hence we can write the output in the form:

$$Y(s) = \frac{1}{6s} + \frac{1}{3(s+3)} - \frac{1}{2(s+2)}$$

Hence, using Table 3.1 gives:

$$y = 0.17 + 0.33\ e^{-3t} - 0.5\ e^{-2t}$$

Problems

1 A system has an input of a voltage of 3 V which is suddenly applied by a switch being closed. What is the input as an s function?

2 A system has an input of a voltage impulse of 2 V. What is the input as an s function?

3 A system has an input of a voltage of a ramp voltage which increases at 5 V per second. What is the input as an s function?

4 A system gives an output of $1/(s + 5)\ V(s)$. What is the output as a function of time?

5 A system has a transfer function of $5/(s + 3)$. What will be its output as a function of time when subject to (a) a unit step input of 1 V, (b) a unit impulse input of 1 V?

6 A system has a transfer function of $2/(s + 1)$. What will be its output as a function of time when subject to (a) a step input of 3 V, (b) an impulse input of 3 V?

7 A system has a transfer function of $1/(s + 2)$. What will be its output as a function of time when subject to (a) a step input of 4 V, (b) a ramp input unit impulse of 1 V/s?

8 Use partial fractions to simplify the following expressions:

(a) $\dfrac{s-6}{(s-1)(s-2)}$, (b) $\dfrac{s+5}{s^2+3s+2}$, (c) $\dfrac{2s-1}{(s+1)^2}$

9 A system has a transfer function of:

$$\frac{8(s+3)(s+8)}{(s+2)(s+4)}$$

What will be the output as a time function when it is subject to a unit step input? Hint: use partial fractions.

10 A system has a transfer function of:

$$G(s) = \frac{8(s+1)}{(s+2)^2}$$

What will be the output from the system when it is subject to a unit impulse input? Hint: use partial fractions.

11 What will be the state of damping of systems having the following transfer functions and subject to a unit step input?

(a) $\dfrac{1}{s^2+2s+1}$, (b) $\dfrac{1}{s^2+7s+12}$, (c) $\dfrac{1}{s^2+s+1}$

12 The input x and output y of a system are described by the differential equation:

$$\frac{dy}{dt} + 2y = x$$

Determine how the output will vary with time when there is an input which starts at zero time and then increases at the constant rate of 6 units/s. The initial output is zero.

13 The input x and output y of a system are described by the differential equation:

$$\frac{d^2y}{dt^2} + 3\frac{dy}{dt} + 2y = x$$

If initially the input and output are zero, what will be the output when there is a unit step input?

14 The input x and output y of a system are described by the differential equation:

$$\frac{d^2y}{dt^2} + 4\frac{dy}{dt} + 3y = x$$

If initially the input and output are zero, what will be the output when there is a unit impulse input?

15 A control system has a forward path transfer function of $2/(s + 2)$ and a negative feedback loop with transfer function 4. What will be the response of the system to a unit step input?

16 A system has a transfer function of $100/(s^2 + s + 100)$. What will be its natural frequency ω_n and its damping ratio ζ?

17 A system has a transfer function of $10/(s^2 + 4s + 9)$. Is the system under-damped, critically damped or over-damped?

18 A system has a transfer function of $3/(s^2 + 6s + 9)$. Is the system under-damped, critically damped or over-damped?

19 A system has a forward path transfer function of $10/(s + 3)$ and a negative feedback loop with transfer function 5. What is the time constant of the resulting first-order system?

4 System parameters

4.1 Introduction

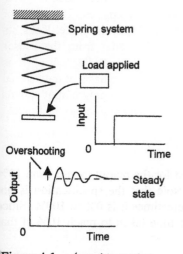

Figure 4.1 *A spring system with an output to a step input which takes time to reach the steady state value and shows overshooting*

When a system is subject to, say, a unit step input it may give an output which eventually settles down to some steady state response. The response that it gives before settling down to this steady state is called its *transient response*. This chapter is about the parameters used to specify the transient response of systems and whether the transients lead to unstable systems.

For example, if we have a spring system (Figure 4.1) and suddenly apply a load to it, it has a transient response which results in it taking some time to reach its steady state value and also it is likely to overshoot the steady state value before it finally settles down to the steady state value. What factors can we change with the spring system in order to get it to respond more quickly to an input and also to minimise the overshooting? These are questions that are often posed for control systems. As another illustration, consider a control system used with an automatic machine to position a workpiece before some machining operation, we need to know how fast the system will respond to an input signal and position the item in the required position and will the system be like the spring system when a load is applied to it and overshooting of the required position occur. Overshooting is undesirable in such a situation and so, if it occurred in such a control system, we would need to consider what steps can be taken to eliminate it. Parameters are used as a way of specifying how fast a system will respond to an input and how quickly it will settle down to its steady state value.

With the above spring system, the result of applying a load is that, after some oscillations with ever decreasing amplitude, the transients die away and the system settles down to a stead state value. The system, is said to be stable. If, however, the oscillations had continued with ever increasing amplitude, then no steady state value would have been reached and the system would be unstable. This chapter takes a brief look at the conditions for stability of systems.

4.2 First order systems

A first order system has a transfer function of the form:

$$G(s) = \frac{k}{\tau s + 1}$$

where k is the steady state gain and τ the time constant. When a unit step input is applied to such a system, the output y is:

$$y = k(1 - e^{-t/\tau})$$

When the time $t = \tau$ then we have $y = k(1 - e^{-1}) = 0.63k$. Thus, the *time constant* τ for a first order system when subject to a step input is the time taken for the output to reach 0.63, of the steady state value.

For a first order system, the parameters used to specify the transient performance are:

1   *Delay time*

The delay time t_d is the time required for the output response to reach 50% of its steady state value. Thus, since k is the final value, the time taken to reach 50% of this value is given by:

$$\tfrac{1}{2}k = k(1 - e^{-t_d/\tau})$$

$$\tfrac{1}{2} = e^{-t_d/\tau}$$

$$\ln 2 = \frac{t_d}{\tau}$$

$$t_d = \tau \ln 2$$

2   *Rise time*

The rise time t_r is the time required for the output to rise from 10% to 90% of its steady state value. Note that the specification is not always in terms of 10% to 90%, sometimes it is 0% to 100%. Since k is the steady state value then the time taken to reach 10% of that value is:

$$\frac{10}{100}k = k(1 - e^{-t_{10}/\tau})$$

$$\frac{1}{10} = e^{-t_{10}/\tau}$$

$$\ln 10 = \frac{t_{10}}{\tau}$$

$$t_{10} = \tau \ln 10$$

The time taken to reach 90% of the steady state value is given by:

$$\frac{90}{100}k = k(1 - e^{-t_{90}/\tau})$$

$$\frac{9}{10} = e^{-t_{90}/\tau}$$

$$\ln 10 - \ln 9 = \frac{t_{90}}{\tau}$$

$$t_{90} = \tau \ln 10 - \tau \ln 9$$

Hence the rise time is:

$$t_r = t_{90} - t_{10} = \tau \ln 9$$

Example

Determine the delay time and the rise time for a first order system with the transfer function:

$$G(s) = \frac{3}{2s + 1}$$

Comparing the transfer function with $k/(\tau s + 1)$ indicates that the steady state gain is 3 and the time constant is 2 s. Thus, the delay time is:

$$t_{10} = \tau \ln 10 = 2 \ln 10 = 4.6 \text{ s}$$

The rise time is:

$$t_r = t_{90} - t_{10} = \tau \ln 9 = 2 \ln 9 = 4.4 \text{ s}$$

Example

A mercury-in-glass thermometer acts as a first order system with an input of temperature and an output of the mercury position against a scale. The thermometer is initially at 0°C and is then suddenly placed in water at 100°C. After 80 s the thermometer reads 98°C. Determine (a) the time constant, (b) the delay time, (c) the rise time.

(a) For such a system the output θ is related to the input by the equation:

$$\theta = 100(1 - e^{-t/\tau})$$

Hence:

$$98 = 100(1 - e^{-80/\tau})$$

$$0.2 = e^{-80/\tau}$$

$$\ln 0.2 = -\frac{80}{\tau}$$

Hence, the time constant τ is 49.7 s.
(b) The delay time is:

$$t_{10} = \tau \ln 10 = 49.7 \ln 10 = 114.4 \text{ s}$$

(c) The rise time is:

$$t_r = t_{90} - t_{10} = \tau \ln 9 = 49.7 \ln 9 = 109.2 \text{ s}$$

4.3 Second order systems   A second order system has a transfer function of the form:

$$G(s) = \frac{k\omega_n^2}{s^2 + 2\zeta\omega_n s + \omega_n^2}$$

The form the output takes for a unit step input depends on whether the damping factor ζ is less than, equal to or greater than 1, the corresponding outcomes being underdamped, critically damped and overdamped (see Section 3.6).

For the underdamped oscillations of a system we have the output y given by:

$$y = k\left[1 - \frac{e^{-\zeta\omega_n t}}{\sqrt{1-\zeta^2}}\sin\left(\omega_n\sqrt{(1-\zeta^2)}\,t + \phi\right)\right]$$

with the damped natural frequency ω given by:

$$\omega = \omega_n\sqrt{(1-\zeta^2)}$$

We can write the above equation for the output in what is often a more convenient form. Since $\sin(A+B) = \sin A \cos B + \cos A \sin B$, the sine term can be written as:

$$\sin(\omega t + \phi) = \sin\omega t \cos\phi + \cos\omega t \sin\phi$$

and since ϕ is a constant:

$$\sin(\omega t + \phi) = P\sin\omega t + Q\cos\omega t$$

where P and Q are constants. Thus the output can be written as:

$$y = k\left[1 - \frac{e^{-\zeta\omega_n t}}{\sqrt{1-\zeta^2}}(P\sin\omega t + Q\cos\omega t)\right]$$

The performance of an underdamped second order system to a unit step input (Figure 4.2) can be specified by:

1   *Rise time*

The rise time t_r is the time taken for the response x to rise from 0 to the steady-state value y_{ss}. This is the time for the oscillating response to complete a quarter of a cycle, i.e. $\frac{1}{2}\pi$. Thus:

$$\omega t_r = \frac{1}{2}\pi$$

We can thus reduce the rise time by increasing the damped natural frequency, this value being determined by the undamped natural angular frequency and the damping factor, i.e.

$$t_r = \frac{\pi}{2\omega_n\sqrt{1-\zeta^2}}$$

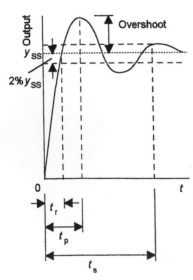

Figure 4.2   *Step response of an under-damped system*

The rise time is sometimes specified as the time taken for the response to rise from 10% to 90% of the steady state value.

2   *Peak time*

The peak time t_p is the time taken for the response to rise from 0 to the first peak value. This is the time for the oscillating response to complete one half-cycle, i.e. π. Thus:

$$\omega t_p = \pi$$

and so we can write:

$$t_p = \frac{\pi}{\omega_n \sqrt{1-\zeta^2}}$$

When ζ is 1 then the peak time becomes infinite; this indicates that at critical damping the steady state value is never reached but only approached asymptotically.

3   *Overshoot*

The overshoot is the maximum amount by which the response overshoots the steady state value and is thus the amplitude of the first peak. The overshoot is often written as a percentage of the steady state value.

The steady state value is when t tends to infinity and thus $y_{ss} = k$. Since $y = 0$ when $t = 0$ then, since $e^0 = 1$, we have:

$$0 = k\left[1 - \frac{1}{\sqrt{1-\zeta^2}}(0+Q)\right]$$

and so $Q = \sqrt{(1-\zeta^2)}$.

The overshoot occurs at $\omega t = \pi$ and thus:

$$y = k\left[1 - \frac{e^{-\zeta \omega_n t}}{\sqrt{1-\zeta^2}}(P\sin\omega t + Q\cos\omega t)\right]$$

becomes:

$$y = y_{ss}\left[1 - \frac{e^{-\zeta \omega_n \pi/\omega}}{\sqrt{1-\zeta^2}}(0-Q)\right]$$

The overshoot is the difference between the output at that time and the steady-state value. Hence:

$$\text{overshoot} = y_{ss}\frac{e^{-\zeta \omega_n \pi/\omega}}{\sqrt{1-\zeta^2}}Q = y_{ss}\,e^{-\zeta \omega_n \pi/\omega}$$

Since $\omega = \omega_n \sqrt{(1-\zeta^2)}$ then we can write:

$$\text{Overshoot} = y_{ss} \exp\left(\frac{-\zeta\omega_n\pi}{\omega_n\sqrt{1-\zeta^2}}\right)$$

$$= y_{ss} \exp\left(\frac{-\zeta\pi}{\sqrt{1-\zeta^2}}\right)$$

Expressed as a percentage of y_{ss}:

$$\text{percentage overshoot} = \exp\left(\frac{-\zeta\pi}{\sqrt{1-\zeta^2}}\right) \times 100\%$$

Note that the overshoot does not depend on the natural frequency of the system but only on the damping factor. As the damping factor approaches 1 so the percentage overshoot approaches zero. Table 4.1 gives values of the percentage overshoot for particular damping ratios.

Table 4.1 *Percentage peak overshoot*

Damping ratio	Percentage overshoot
0.2	52.7
0.3	37.2
0.4	25.4
0.5	16.3
0.6	9.5
0.7	4.6
0.8	1.5
0.9	0.2

4   *Subsidence ratio*
An indication of how fast oscillations decay is provided by the *subsidence ratio* or *decrement*. This is the amplitude of the second overshoot divided by the amplitude of the first overshoot. The first overshoot occurs when we have $\omega t = \pi$ and so:

$$\text{first overshoot} = y_{ss} \exp\left(\frac{-\zeta\pi}{\sqrt{1-\zeta^2}}\right)$$

The second overshoot occurs when $\omega t = 3\pi$ and so:

$$\text{second overshoot} = y_{ss} \exp\left(\frac{-3\zeta\pi}{\sqrt{1-\zeta^2}}\right)$$

Thus the subsidence ratio is given by:

$$\text{subsidence ratio} = \frac{\text{second overshoot}}{\text{first overshoot}} = \exp\left(\frac{-2\zeta\pi}{\sqrt{1-\zeta^2}}\right)$$

5   *Settling time*

The settling time t_s is used as a measure of the time taken for the oscillations to die away. It is the time taken for the response to fall within and remain within some specified percentage of the steady-state value (see Figure 4.1). Thus for the 2% settling time, the amplitude of the oscillation should fall to be less than 2% of y_{ss}. We have:

$$y = k\left[1 - \frac{e^{-\zeta\omega_n t}}{\sqrt{1-\zeta^2}}(P\sin\omega t + Q\cos\omega t)\right]$$

with $y_{ss} = k$, $\omega = \omega_n\sqrt{(1-\zeta^2)}$ and, as derived earlier in item 3, $Q = \sqrt{(1-\zeta^2)}$. The amplitude of the oscillation is $(y - y_{ss})$ when y is a maximum value. The maximum values occur when ωt is some multiple of π and thus we have $\cos\omega t = 1$ and $\sin\omega t = 0$. For the 2% settling time, the settling time t_s is when the maximum amplitude is 2% of y_{ss}, i.e. $0.02y_{ss}$. Thus:

$$0.02y_{ss} = y_{ss}\,e^{-\zeta\omega_n t_s}$$

Taking logarithms gives $\ln 0.02 = -\zeta\omega_n t_s$ and since $\ln 0.02 = -3.9$ or approximately 4 then:

$$t_s = \frac{4}{\zeta\omega_n}$$

The above is the value of the settling time if the specified percentage is 2%. If the specified percentage is 5% the equation becomes

$$t_s = \frac{3}{\zeta\omega_n}$$

6   *Number of oscillations to settling time*

The time taken to complete one cycle, i.e. the periodic time, is $1/f$, where f is the frequency, and since $\omega = 2\pi f$ then the time to complete one cycle is $2\pi/f$. In a settling time of t_s the number of oscillations that occur is:

$$\text{number of oscillations} = \frac{\text{settling time}}{\text{periodic time}}$$

and thus for a settling time defined for 2% of the steady-state value:

$$\text{number of oscillations} = \frac{4/\zeta\omega_n}{2\pi/\omega}$$

Since $\omega = \omega_n\sqrt{(1-\zeta^2)}$, then:

$$\text{number of oscillations} = \frac{2\omega_n \sqrt{1-\zeta^2}}{\pi\zeta\omega_n} = \frac{2}{\pi}\sqrt{\frac{1}{\zeta^2}-1}$$

In designing a system the following are the typical points that are considered:

1   For a rapid response, i.e. small rise time, the natural frequency must be large. Figure 4.3 shows the types of response obtained to a unit step input to systems having the same damping factor of 0.2 but different natural angular frequencies. The response time with the natural angular frequency of 10 rad/s, damped frequency 9.7 rad/s, is much higher than that with a natural angular frequency of 1 rad/s, damped frequency 0.97 rad/s.

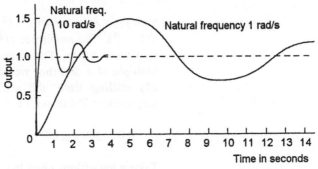

Figure 4.3   *Response of a unit gain second order system to a unit step input, the damping factor being the same for both responses*

2   The damping factor is typically in the range 0.4 to 0.8 since smaller values give an excessive overshoot and a large number of oscillations before the system settles down. Larger values render the system sluggish since they increase the response time. Though, in some systems where no overshoot can be tolerated, a high value of damping factor may have to be used. Figure 4.4 shows the effect on the response of a second order system of a change of damping factor when the natural angular frequency remains unchanged.

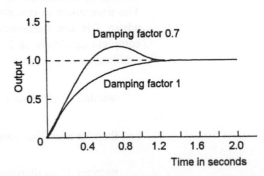

Figure 4.4   *Response of a unit gain second order system to a unit step input, the natural angular frequency being the same for both responses*

Example

A second order system has a natural angular frequency of 2.0 rad/s and a damped frequency of 1.8 rad/s. What are its (a) damping factor, (b) 100% rise time, (c) percentage overshoot, (c) 2% settling time, and (d) the number of oscillations within the 2% settling time?

(a) Since $\omega = \omega_n \sqrt{(1 - \zeta^2)}$, then the damping factor is given by:

$$1.8 = 2.0\sqrt{1 - \zeta^2}$$

and $\zeta = 0.44$.

(b) Since $\omega t_r = \frac{1}{2}\pi$, then the 100% rise time is given by

$$t_r = \frac{\pi}{2 \times 1.8} = 0.87 \text{ s}$$

(c) The percentage overshoot is given by:

$$\% \text{ overshoot} = \exp\left(\frac{-\zeta\pi}{\sqrt{1-\zeta^2}}\right) \times 100\%$$

$$= \exp\left(\frac{-0.44\pi}{\sqrt{1-0.44^2}}\right) \times 100\% = 21\%$$

(d) The 2% settling time is given by:

$$t_s = \frac{4}{\zeta\omega_n} = \frac{4}{0.44 \times 2.0} = 4.5 \text{ s}$$

(e) The number of oscillations occurring within the 2% settling time is given by:

$$\text{number of oscillations} = \frac{2}{\pi}\sqrt{\frac{1}{\zeta^2} - 1} = \frac{2}{\pi}\sqrt{\frac{1}{0.44^2} - 1} = 1.3$$

Example

The feedback system shown in Figure 4.5 has the transfer function of the forward path as $K/[s(s + a)]$ and the transfer function of the feedback path as 1. What will be the effect on the system response of changing the gain K?

The transfer function of the closed loop system is:

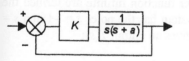

Figure 4.5 *Example*

$$G(s) = \frac{K/[s(s+a)]}{1 + K/[s(s+a)]} = \frac{K}{s(s+a) + K} = \frac{K}{s^2 + as + K}$$

By comparison with the standard form of the transfer function for a second order system, we have $\omega_n = \sqrt{K}$ and $\zeta = a/2\sqrt{K}$. Thus the rise time is given by $\omega t_r = \frac{1}{2}\pi$, with $\omega = \omega_n \sqrt{(1 - \zeta^2)}$, and so is:

$$t_r = \frac{\pi}{2\sqrt{K}\sqrt{1-(a^2/4K)}}$$

The rise time thus decreases as K increases. Thus, increasing the gain decreases the rise time and so increases the speed of response of the system.

The percentage overshoot is given by:

$$\% \text{ overshoot} = \exp\left(\frac{-\zeta\pi}{\sqrt{1-\zeta^2}}\right) \times 100\%$$

$$= \exp\left(\frac{-(a/2\sqrt{K})\pi}{\sqrt{1-(a^2/4K)}}\right) \times 100\%$$

$$= \exp\left(\frac{a\pi}{\sqrt{4K-a^2}}\right) \times 100\%$$

Thus, increasing K results in an increase in the percentage overshoot.

The 2% settling time is given by:

$$t_s = \frac{4}{\zeta\omega_n} = \frac{4}{a/2} = \frac{8}{a}$$

Thus, the settling time is independent of the gain K.

4.4 Stability   We can define a system as being a *stable system* if, when given an input or a change in input, it has transients which die away with time and leave the system in a steady state condition. The system would be *unstable* if the transients did not die away with time but grew with time and so steady state conditions were never reached.

Consider a second order system with the transfer function:

$$G(s) = \frac{1}{(s+1)(s+2)}$$

The values of s which make the transfer function infinite are termed the *poles* of the system. Thus, the above system has the poles $s = -1$ and $s = -2$. A unit step input to such a system gives an output $Y(s)$:

$$Y(s) = \frac{1}{s(s+1)(s+2)} = \frac{1}{2s} - \frac{1}{s+1} + \frac{1}{2(s+2)}$$

This varies with time as:

$$y = \tfrac{1}{2} - e^{-t} + \tfrac{1}{2}e^{-2t}$$

Each of the poles gives rise to a transient term. Both the resulting exponential terms die away with time to give a steady state value of 0.5, the more negative the value of s for a pole the more rapidly the corresponding term dies away. Thus, the system is stable.

Now consider a second order system with the transfer function:

$$G(s) = \frac{1}{(s-1)(s-2)}$$

This system has the poles $s = +1$ and $s = +2$. A unit step input to such a system gives an output $Y(s)$:

$$Y(s) = \frac{1}{s(s-1)(s-2)} = \frac{1}{2s} - \frac{1}{s-1} + \frac{1}{2(s-2)}$$

This varies with time as:

$$y = \tfrac{1}{2} - e^{+t} + \tfrac{1}{2}e^{+2t}$$

Each of the positive poles gives rise to exponential terms which grow with time, the larger the value of s for a pole the more rapidly the correspondiong term grows. Thus, the transients do not die away but increase and so the system is unstable.

In general, if a system has a transfer function with a pole which is negative then it gives rise to a transient which dies away with time, whereas if it has a pole which is positive then the transient grows with time. Thus, if a transfer function has a pole which is positive then it is unstable.

In general, for a second order system we have the transfer function (see Section 3.5):

$$G(s) = \frac{k\omega_n^2}{s^2 + 2\zeta\omega_n s + \omega_n^2} = \frac{k\omega_n^2}{(s+p_1)(s+p_2)}$$

The roots of the quadratic denominator, i.e. the poles, are given by:

$$p_1 = -\zeta\omega_n + \omega_n\sqrt{\zeta^2 - 1}$$

$$p_2 = -\zeta\omega_n - \omega_n\sqrt{\zeta^2 - 1}$$

With $\zeta > 1$ we have real roots then the square root is of a positive quantity and thus the overall root can be written as a real number, as in the examples given above when we had $s = +1$, $s = +2$, $s = -1$ and $s = -2$. With $\zeta < 1$ then the square root is of a negative quantity. If we write j for the square root of minus 1 then the roots can be written as:

$$p_1 = -\zeta\omega_n + j\omega_n\sqrt{1 - \zeta^2}$$

$$p_2 = -\zeta\omega_n - j\omega_n\sqrt{1 - \zeta^2}$$

We can thus write the values of the roots, and so poles, in the form $a + jb$; the jb part of the value is known as an imaginary number. Such

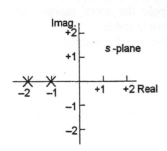

Figure 4.6   *Poles at −1 and −2 for a stable system*

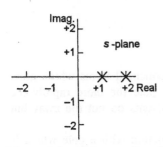

Figure 4.7   *Poles at +1 and +2 for an unstable system*

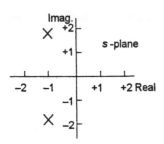

Figure 4.8   *Poles at −1 ± j1.73 for a stable system*

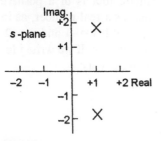

Figure 4.9   *Poles at +1 ± j1.73 for an unstable system*

values give rise to oscillatory transients. However, the same rule applies for stability, namely that if the *a* term is negative then the system is stable and if it is positive it is unstable.

4.4.1 The *s* plane

We can plot the positions of the poles on a graph with the real part of the pole value as the *x*-axis and the imaginary part as the *y*-axis. The resulting graph describes what is termed the *s-plane*.

As an illustration, Figure 4.6 shows the *s*-plane for the transfer function:

$$G(s) = \frac{1}{(s+1)(s+2)}$$

with $s = -1$ and $s = -2$, there being no imaginary terms. This describes an overdamped system. Figure 4.7 shows the *s*-plane for the transfer function:

$$G(s) = \frac{1}{(s-1)(s-2)}$$

with $s = +1$ and $s = +2$, there being no imaginary terms. For the transfer function:

$$G(s) = \frac{1}{s^2 + 2s + 4}$$

we have the roots of the quadratic given by:

$$s = \frac{-2 \pm \sqrt{4-16}}{2} = -1 \pm j1.73$$

Figure 4.8 shows the *s*-plane for this transfer function, this describing an underdamped system. For the transfer function:

$$G(s) = \frac{1}{s^2 - 2s + 4}$$

we have the roots of the quadratic given by:

$$s = \frac{2 \pm \sqrt{4-16}}{2} = +1 \pm j1.73$$

Figure 4.9 shows the *s*-plane for this transfer function. For the transfer function:

$$G(s) = \frac{1}{(s+1)^2}$$

we have the roots $s = -1$ and $s = -1$. This is critical damping. Figure 4.10 shows the *s*-plane for this transfer function.

In general we can state (Figure 4.11):

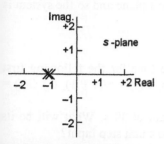

Figure 4.10 *Double pole at −1 for critical damping*

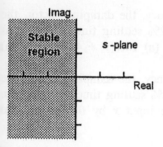

Figure 4.11 *The s-plane: stability when poles are in the left half*

A system is stable if all the system poles lie in the left half of the s-plane.

The relationship between the location of a pole and the form of transient is shown in Figure. 4.12. The more negative the real part of the pole the more rapidly the transient dies away. The larger the imaginary part of the pole the higher the frequency of the oscillation. A system having a pole which has a positive real part is unstable.

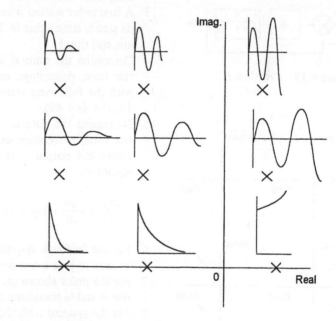

Figure 4.12 *Relationship between pole location and the resulting transient: each oscillatory transient arises from a pair of roots a ± jb with only one of them shown in the figure*

Example

Which of the following systems are stable: (a) $G(s) = 1/(s^2 + s + 1)$, (b) $G(s) = 1/(s^2 - 5s + 4)$, (c) $G(s) = 1/(s^2 - 2s + 3)$?

(a) This has poles of :

$$s = \frac{-1 \pm \sqrt{1-4}}{2} = -0.5 \pm j1.73$$

The poles will lie in the left half of the s-plane and so the system is stable.

(b) This has poles of $s = +4$ and $s = +1$ and so the poles lie in the right half of the s-plane and the system is unstable.

(c) This has poles of:

$$s = \frac{2 \pm \sqrt{4-12}}{2} = +1 \pm j2.8$$

The poles will lie in the right half of the *s*-plane and so the system is unstable.

Problems

1 Determine the delay time and the rise time for the following first order systems: (a) $G(s) = 1/(4s + 1)$, (b) $G(s) = 5/(s + 1)$, (c) $G(s) = 2/(s + 3)$.

2 A first order system has a time constant of 30 s. What will be its delay time and rise time when subject to a unit step input?

3 A first order system when subject to a unit step input rises to 90% of its steady state value in 20 s. Determine its time constant, delay time and rise time?

4 Determine the natural angular frequency, the damping factor, the rise time, percentage overshoot and 2% settling time for systems with the following transfer functions: (a) $100/(s^2 + 4s + 100)$, (b) $49/(s^2 + 4s + 49)$.

5 Determine the natural angular frequency, the damping factor, the rise time, percentage overshoot and 2% settling time for a system where the output *y* is related to the input *x* by the differential equation:

$$\frac{d^2y}{dt^2} + 5\frac{dy}{dt} + 16y = 16x$$

6 For the feedback system shown in Figure 4.13, what gain *K* should be used to give a rise time of 2 s?

7 For the poles shown on the *s*-planes in Figure 4.14, which will give rise to stable transients and which to oscillating transients?

8 Are the systems with the following transfer functions stable?

(a) $\dfrac{1}{s^2 + 2s + 1}$, (b) $\dfrac{1}{s^2 - 2s + 10}$, (c) $\dfrac{1}{(s+1)(s-3)}$

9 Figure 4.15 shows a feedback control system with unity feedback. Will the system be stable when (a) $K = 1$, $G(s) = 1/[s(s + 1)]$, (b) $K = 3$, $G(s) = 1/[s + 4)(s - 1)]$, (c) $K = 5$, $G(s) = 1/[s + 4)(s - 1)]$?

10 State if the following systems are stable, the relationship between input *x* and output *y* being described by the differential equations :

(a) $\dfrac{d^2y}{dt^2} + 3\dfrac{dy}{dt} + 2y = x$, (b) $\dfrac{d^2y}{dt^2} + \dfrac{dy}{dt} - 6y = x$

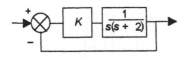

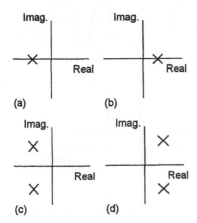

Figure 4.13   *Problem 6*

Figure 4.14   *Problem 7*

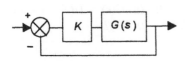

Figure 4.15   *Problem 9*

5 Frequency response

5.1 Introduction

In earlier chapters we have considered the outputs that arise from systems when subject to step, impulse and ramp inputs. In this chapter we consider the steady state output responses of systems when the inputs are sinusoidal signals. This leads to powerful methods of analysing systems in considering how the amplitude and phase of the output signal varies as the frequency of the input sinusoidal signal is changed. This variation is termed the *frequency response* of the system and can be described by, what are termed, *Bode diagrams*.

The derivation of a transfer function for system involves making assumptions about the physical model, e.g. springs, masses and damping, that can be used to represent the system. We can derive the frequency response of systems from a knowledge of their transfer functions. Thus the validity of the resulting transfer function can be tested by experiment using sinusoidal inputs and comparing the experimental frequency response with that which is predicted by the transfer function. Conversely, we can determine the frequency response for a system and then use it to predict the form of the transfer function.

Thus, this chapter shows how frequency response information can be obtained from the transfer function and how it can be presented graphically by means of a Bode diagram. The use of experimentally determined Bode plots to estimate the transfer functions of systems is then discussed. The chapter concludes with a discussion of the parameters used to describe the stability of systems and their determination from Bode plots, also compensation techniques which can be used to enhance the stability of systems.

5.2 Phasors

In discussing sinusoidal signals, a convenient way of representing such signals is by *phasors*. We imagine a sinusoidal signal $y = Y \sin \omega t$, i.e. amplitude Y and angular frequency ω, being produced by a radial line of length Y rotating with a constant angular velocity ω (Figure 5.1) and taking the vertical projection y of the line at any instant of time to represent the value of the sinusoidal signal. If we have another sinusoidal signal of different amplitude then the radial line will be of a different length. If we have a sinusoidal signal with a different phase then it will start with a different value at time $t = 0$ and so the radial line will start at $t = 0$ at some angle, termed the *phase angle*, to the reference axis. The reference axis is usually taken as the horizontal axis. Such lines are termed *phasors* and the representation is said to be in the *frequency-domain*.

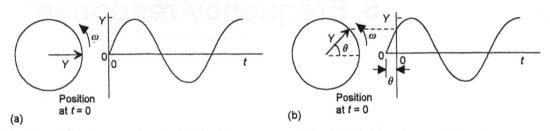

Figure 5.1 *Sinusoidal signals represented by rotating lines, i.e. phasors: (a) y = Y sin ωt, (b) y = Y sin(ωt + θ)*

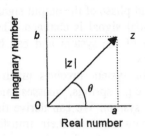

Figure 5.2   *Representing a complex number by a line on an Argand diagram*

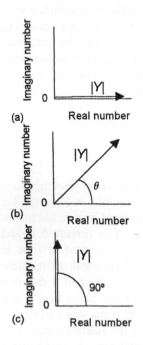

Figure 5.3   *Representing phasors by lines on an Argand diagram: (a) no phase angle, (b) phase θ, (c) phase 90°*

When we use a phasor to describe a sinusoidal signal, all we need specify is its magnitude and phase angle.

In order to clearly indicate when we are talking of the magnitude of a sinusoidal signal we often write $|Y|$ for the magnitude of the sinusoidal signal represented by the phasor and bold, non-italic, letters for the symbols for phasors, e.g. **Y**. Thus, **Y** implies a phasor with both magnitude and phase.

A complex number $z = a + jb$ can be represented on an Argand diagram, i.e. a graph of imaginary part plotted against real part, by a line (Figure 5.2) of length $|z|$ at an angle $θ$. The magnitude $|z|$ of a complex number z and its angle $θ$ is thus given by:

$$|z| = \sqrt{a^2 + b^2}$$

$$θ = \tan^{-1}\left(\frac{b}{a}\right)$$

We can describe a phasor used to represent a sinusoidal quantity by a complex number. Thus, if we have $y = Y \sin ωt$ then this is described by a phasor (Figure 5.3(a)) consisting of just a real number. Thus, a unit magnitude phasor with phase angle 0° is represented by $1 + j0$. However, for $y = Y \sin (ωt + θ)$ we have a phasor (Figure 5.3(b)) which has, in general, both a real and imaginary part and so is represented by $a + jb$. If the phase $θ$ is 90° then for $y = \sin (ωt + 90°) = \cos ωt$ the phasor (Figure 5.3(c)) has only an imaginary part. Thus, such a unit magnitude phasor is represented by $0 + j1$.

If we have a phasor of length 1 and phase angle 0° (Figure 5.4(a)) then it will have a complex representation of $1 + j0$. The same length phasor with a phase angle of 90° (Figure 5.4(b)) will have a complex representation of $0 + j1$; rotation of a phasor anticlockwise by 90° corresponds to multiplication of the phasor by j since $j(1 + j0) = 0 + j1$. If we now rotate this phasor by a further 90° (Figure 5.4(c)), then as $j(0 + j1) = 0 + j^2 1$ we have the original phasor multiplied by j^2. As this phasor is just the original phasor in the opposite direction, it is just the original phasor multiplied by -1 and so $j^2 = -1$ and hence $j = \sqrt{(-1)}$. Rotation of the original phasor through a total of 270°, i.e. $3 \times 90°$, is equivalent to multiplying the original phasor by $j^3 = j(j^2) = -j$.

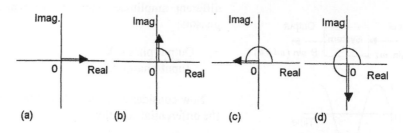

Figure 5.4 *Phasor rotated by (a) 0°, (b) 90°, (c) 180°, (d) 270°*

Example

What magnitude and phase is given the phasors described as (a) j3, (b) 1 + j2?

(a) The magnitude is 3 units and, since we only have an imaginary component, the phase is 90°.
(b) The magnitude is $\sqrt{(a^2 + b^2)} = \sqrt{(1 + 4)} = 2.2$ units and the phase is given by $\tan\theta = b/a = 2/1$ and $\theta = 63.4°$.

5.3 Sinusoidal inputs

If we consider an electrical system, such as an electric circuit involving a resistor in series with a capacitor (Figure 5.5), then when we have a sinusoidal signal input we obtain a steady state output, e.g. the potential difference across the capacitor in Figure 5.1, which is also sinusoidal and has the same frequency as the input. The output can, however, have a different amplitude to the input and be shifted in phase from it. For Figure 5.1, the p.d. across the capacitor leads the input by a phase angle ϕ given by $\tan\phi = X_C/R$; the capacitive reactance $X_C = \omega C$. The amplitude of the output $V_C = IX_C$, where I is the amplitude of the circuit current. The amplitude and phase depend on the frequency of the input.

This type of relationship between the steady state output and a sinusoidal input applies to all forms of system. Thus, in general we can describe systems in the way shown in Figure 5.6 and so:

The frequency response of a system is the steady state response of the system to a sinusoidal input signal. The steady state output is a sinusoidal signal of the same frequency as the input signal, differing only in amplitude and phase angle.

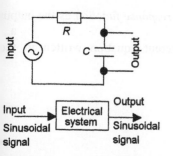

Figure 5.5 *An electrical system with a sinusoidal input*

5.3.1 Frequency response function

In order to arrive at the principle of the frequency response function we will consider a simple system with a sinusoidal input and steady-state sinusoidal output and recognise that our conclusions can be applied in a more general way to all systems.

Consider a system where the input x is related to the output by $y = kx$. If we have an input of a sinusoidal signal $x = \sin\omega t$ then the output is $y = k\sin\omega t$ and so a sinusoidal signal with the same frequency but a

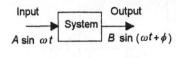

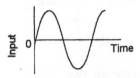

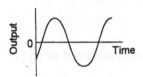

Figure 5.6   *The frequency response of a system*

different amplitude. Thus, if we represent the sinusoidal signals by phasors:

$$\frac{\text{Output phasor } \mathbf{Y}}{\text{Input phasor } \mathbf{X}} = k$$

Now consider a system where the input x is related to the output y by the differential equation:

$$\tau\frac{dy}{dt} + y = kx$$

Thus, since the frequency does not change we can take $x = \sin \omega t$ and $y = \sin \omega t$ and so, since $dy/dt = \omega \cos \omega t = \omega \sin (\omega t + 90°)$, the equation can be written as:

$$\tau\omega \sin (\omega t + 90°) + \sin \omega t = k \sin \omega t$$

We can represent sinusoidal signals by phasors and describe them by complex numbers. Thus, the above equation can be written in terms of phasors as:

$$j\tau\omega\mathbf{Y} + \mathbf{Y} = k\mathbf{X}$$

Hence, we can write:

$$\frac{\text{output phasor } \mathbf{Y}}{\text{input phasor } \mathbf{X}} = \frac{k}{1 + j\omega\tau}$$

This leads to a definition of a *frequency response function* as the output phasor divided by the input phasor.

We can compare this with the different equation written in the s-domain as:

$$\tau sY(s) + Y(s) = kX(s)$$

and the resulting transfer function:

$$G(s) = \frac{Y(s)}{X(s)} = \frac{k}{1 + \tau s}$$

The frequency response function equation is of the same form as the transfer function if we replace s by $j\omega$. Hence the frequency response function is denoted by $G(j\omega)$. In general we can state:

The frequency response function is obtained from the transfer function by replacing s by jω.

Example

Determine the frequency response function for a system having a transfer function of $G(s) = 5/(2 + s)$.

Replacing s by $j\omega$ gives the frequency response function of $G(j\omega) = 5/(2 + j\omega)$.

5.3.2 Frequency response for first-order systems

Consider a first-order system and the determination, from the frequency response function, of the magnitude and phase of the steady-state output when it is subject to a sinusoidal input. For example, we might have a system which can be represented as a capacitor in series with a resistor and consider the output p.d. across the capacitor when there is a sinusoidal voltage input.

In general, a first-order system has a transfer function of the form:

$$G(s) = \frac{1}{1 + \tau s}$$

where τ is the time constant of the system. The frequency response function $G(j\omega)$ can be obtained by replacing s by $j\omega$. Hence:

$$G(j\omega) = \frac{1}{1 + j\omega\tau}$$

We can put this into the form $a + jb$ by multiplying the top and bottom of the expression by $(1 - j\omega\tau)$ to give:

$$G(j\omega) = \frac{1}{1 + j\omega\tau} \times \frac{1 - j\omega\tau}{1 - j\omega\tau} = \frac{1 - j\omega\tau}{1 + j^2\omega^2\tau^2}$$

But $j^2 = -1$, thus we can write this equation as:

$$G(j\omega) = \frac{1}{1 + \omega^2\tau^2} - j\frac{\omega\tau}{1 + \omega^2\tau^2}$$

The frequency response function has thus a real element of $1/(1 + \omega^2\tau^2)$ and an imaginary element of $-\omega\tau/(1 + \omega^2\tau^2)$. Since $G(j\omega)$ is the output phasor divided by the input phasor, then the output phasor has a magnitude bigger than that of the input phasor by a factor $|G(j\omega)|$ given by $\sqrt{(a^2 + b^2)}$ as:

$$|G(j\omega)| = \sqrt{\left(\frac{1}{1 + \omega^2\tau^2}\right)^2 + \left(\frac{\omega\tau}{1 + \omega^2\tau^2}\right)^2}$$

$$= \frac{1}{\sqrt{1 + \omega^2\tau^2}}$$

$|G(j\omega)|$ is referred to as the *gain* of the system.

The phase difference ϕ between the output phasor and the input phasor is given by $\tan \phi = b/a$ as:

$$\tan \phi = -\omega\tau$$

The negative sign indicates that the output phasor lags behind the input phasor by this angle. Thus:

The gain and phase of a system when subject to a sinusoidal input is obtained by putting the frequency response function in the form $a + jb$ and then the gain is $\sqrt{(a^2 + b^2)}$ and the phase is $\tan^{-1} (b/a)$.

Example

Determine the magnitude and phase of the output from a system when subject to a sinusoidal input of $2 \sin 3t$ if it has a transfer function of $G(s) = 2/(s + 1)$.

The frequency-response function is obtained by replacing s by $j\omega$:

$$G(j\omega) = \frac{2}{j\omega + 1}$$

Multiplying top and bottom of the equation by $(-j\omega + 1)$ gives:

$$G(j\omega) = \frac{-j2\omega + 2}{\omega^2 + 1} = \frac{2}{\omega^2 + 1} - j \frac{2\omega}{\omega^2 + 1}$$

The magnitude is thus:

$$|G(j\omega)| = \sqrt{\frac{2^2}{(\omega^2 + 1)^2} + \frac{2^2\omega^2}{(\omega^2 + 1)^2}}$$

$$= \frac{2}{\sqrt{\omega^2 + 1}}$$

and the phase angle is given by:

$$\tan \phi = -\omega$$

For the specified input we have $\omega = 3$ rad/s. The magnitude is thus:

$$|G(j\omega)| = \frac{2}{\sqrt{3^2 + 1}} = 0.63$$

and the phase is given by $\tan \phi = -3$ as $\phi = -72°$. This is the phase angle between the input and the output. Thus, the output is the sinusoidal signal of the same frequency as the input signal and described by $1.26 \sin (3t - 72°)$.

5.3.3 Frequency response for second-order systems

Consider a second-order system and the determination, from the frequency response function, of the magnitude and phase of the steady-state output when it is subject to a sinusoidal input. For example, we might have a system which can be represented as an inductor, a capacitor and a resistor all in series and consider the output p.d. across the capacitor when there is a sinusoidal voltage input.

In general, a second-order system has a transfer function of the form:

$$G(s) = \frac{\omega_n^2}{s^2 + 2\zeta\omega_n s + \omega_n^2}$$

where ω_n is the natural angular frequency and ζ the damping ratio. The frequency-response function is obtained by replacing s by $j\omega$. Thus:

$$G(j\omega) = \frac{\omega_n^2}{-\omega^2 + j2\zeta\omega\omega_n + \omega_n^2} = \frac{\omega_n^2}{(\omega_n^2 - \omega^2) + j2\zeta\omega_n}$$

$$= \frac{1}{\left[1 - \left(\frac{\omega}{\omega_n}\right)^2\right] + j2\zeta\left(\frac{\omega}{\omega_n}\right)}$$

Multiplying the top and bottom of the expression by:

$$\left[1 - \left(\frac{\omega}{\omega_n}\right)^2\right] - j2\zeta\left(\frac{\omega}{\omega_n}\right)$$

gives:

$$G(j\omega) = \frac{\left[1 - \left(\frac{\omega}{\omega_n}\right)^2\right] - j2\zeta\left(\frac{\omega}{\omega_n}\right)}{\left[1 - \left(\frac{\omega}{\omega_n}\right)^2\right]^2 + \left[2\zeta\left(\frac{\omega}{\omega_n}\right)\right]^2}$$

This is of the form $a + jb$ and so, since $G(j\omega)$ is the output phasor divided by the input phasor, we have the magnitude of the output phasor bigger than that of the input phasor, i.e. the gain, by a factor:

$$|G(j\omega)| = \frac{1}{\sqrt{\left[1 - \left(\frac{\omega}{\omega_n}\right)^2\right]^2 + \left[2\zeta\left(\frac{\omega}{\omega_n}\right)\right]^2}}$$

The phase ϕ difference between the input and output is given by:

$$\tan\phi = -\frac{2\zeta\left(\frac{\omega}{\omega_n}\right)}{1 - \left(\frac{\omega}{\omega_n}\right)^2}$$

The minus sign indicates that the output phasor lags the input phasor.

5.4 Bode plots

The frequency response of a system is described by values of the gain and the phase angle which occur when the sinusoidal input signal is varied over a range of frequencies. The term *Bode plot* is used for the pair of graphs of the logarithm to base 10 of the gain plotted against the logarithm to base 10 of the frequency and the phase angle plotted against the logarithm to base 10 of the frequency. The reason for the graphs being in this form is that it enables plots for complex frequency response functions to be obtained by merely adding together the plots obtained for each constituent element.

Suppose we have a sinusoidal input to a number of systems in series (Figure 5.7). The first produces a gain of $|G_1(j\omega)|$ and a phase angle shift of ϕ_1, the second produces a gain of $|G_2(j\omega)|$ and a phase angle shift of ϕ_2, and the third produces a gain of $|G_3(j\omega)|$ and a phase angle shift of ϕ_3. The overall gain of the system will be the products of the gains of each of the systems and thus be:

$$|G(j\omega)| = |G_1(j\omega)||G_2(j\omega)||G_3(j\omega)|$$

and the total phase shift will be:

$$\phi = \phi_1 + \phi_2 + \phi_3$$

The total phase is thus the sum of the phases of the individual elements. If we take the logarithms of the gain equation we obtain:

$$\lg |G(j\omega)| = \lg |G_1(j\omega)| + \lg |G_2(j\omega)| + \lg |G_3(j\omega)|$$

Thus, when we use the logarithms of the gains, we obtain the overall gain by just adding the logarithms of the gains of the individual elements. This enables us to consider any frequency response function as being made up of a number of simple elements and so obtain the response by adding the logarithms of the gains of the simple elements. For example, we can think of the frequency response function $5/(2 + j\omega)$ as being two elements, one with frequency response function 5 and the other with frequency response function $1/(2 + j\omega)$.

Because we can obtain the Bode plot for a system by considering the plots for the constituent elements of its frequency response function and adding, it is only necessary to know the form of the Bode plots for a small number of transfer function terms:

1   Constants
2   $1/s$
3   s
4   $1/(\tau s + 1)$
5   $(\tau s + 1)$
4   $\omega_n^2/(s^2 + 2\zeta\omega_n s + \omega_n^2)$
5   $(s^2 + 2\zeta\omega_n s + \omega_n^2)/\omega_n^2$

The following sections show these Bode plots.

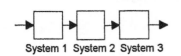

System 1   System 2   System 3

Figure 5.7   *Systems in series*

It is usual to express the gain in decibels (dB), with:

$$\text{gain in dB} = 20 \lg |G(j\omega)|$$

Thus, a gain of 3 is a gain in dB of $20 \lg 3 = 9.5$ dB; a gain of 10 is a gain in dB of $20 \lg 10 = 20$ dB.

5.4.1 Transfer function a constant

This has $G(s) = K$ and so $G(j\omega) = K$. The gain is $|G(j\omega)| = K$ and in decibels this is $20 \lg K$ dB. The phase $= 0°$ if K is positive and $-180°$ if K is negative. The form of the Bode plot is shown in Figure 5.8. Thus, for $K = 10$ the gain is a constant line of $20 \lg 10 = 20$ dB and the phase a constant line of $0°$.

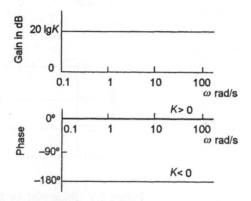

Figure 5.8 *Bode plot for constant transfer function*

5.4.2 Transfer function 1/s

For $G(s) = 1/s$, i.e. on the s-plane (see Section 4.4) the pole is at the origin, we have the frequency response function of $G(j\omega) = 1/(j\omega) = -(j/\omega)$. For such a system the gain is:

$$|G(j\omega)| = 20 \lg(1/\omega) = -20 \lg \omega \text{ dB}$$

and the phase $-90°$. The plots are thus straight lines. When $\omega = 1$ rad/s then the gain is 0. When $\omega = 10$ rad/s it is -20 dB. When $\omega = 100$ rad/s it is -40 dB. Thus the slope of the gain line for a transfer function of $1/s$ is -20 dB for each tenfold increase in frequency (this is termed a decade).

For $G(s) = 1/s^n$ we have $G(j\omega) = 1/(j\omega)^n = -(j/\omega)^n$. For such a system the gain is:

$$|G(j\omega)| = 20 \lg(1/\omega)^n = -20n \lg \omega \text{ dB}$$

and the phase $-n90°$. When $\omega = 1$ rad/s then the magnitude is 0. When $\omega = 10$ rad/s it is $-20n$ dB. When $\omega = 100$ rad/s it is $-40n$ dB. Thus the slope of the gain line for a transfer function of $1/s^n$ is $-20n$ for each tenfold increase in frequency. Figure 5.9 shows the Bode plot.

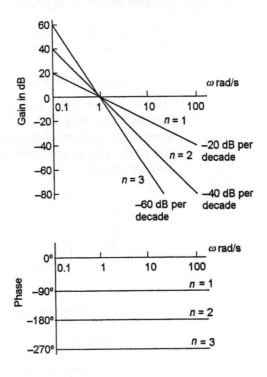

Figure 5.9 *Bode plot for transfer functions having 1/s terms*

5.4.3 Transfer function s

For a transfer function $G(s) = s$ (note that values of s which make the numerator of a transfer function zero are termed zeros, the values which make the denominator zero being termed poles, and so $G(s) = s$ is for a zero at the origin) we have a frequency response function of $G(j\omega) = (j\omega)$. For such a system the gain is:

$$|G(j\omega)| = 20 \lg \omega = 20 \lg \omega \text{ dB}$$

and the phase 90°. The Bode plots are thus straight lines. When $\omega = 1$ rad/s then the gain is 0. When $\omega = 10$ rad/s it is 20 dB. When $\omega = 100$ rad/s it is 40 dB. Thus the slope of the line is 20 dB for each tenfold increase in frequency.

For a transfer function $G(s) = s^n$ and so a frequency response function $G(j\omega) = (j\omega)^m$, the gain is:

$$|G(j\omega)| = 20 \lg \omega^m = 20m \lg \omega \text{ dB}$$

and the phase $m90°$. When $\omega = 1$ rad/s then the magnitude is 0; when $\omega = 10$ rad/s it is $20m$ dB. Thus the slope of the line is $20m$ dB for each tenfold increase in frequency. Figure 5.10 shows the Bode plot.

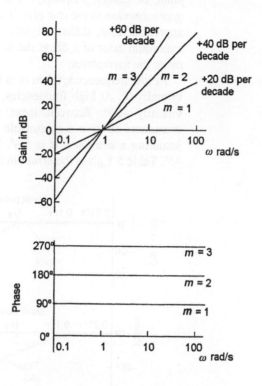

Figure 5.10   *Bode plot for transfer functions s^m*

5.4.4 Transfer function $1/(1 + \tau s)$

For $G(s) = 1/(1 + \tau s)$, i.e. a pole at $s = -1/\tau$ where τ is the time constant, the frequency response is:

$$G(j\omega) = \frac{1}{1 + j\omega\tau} = \frac{1 - j\omega\tau}{1 + \omega^2\tau^2}$$

This means a gain in dB of:

$$\text{gain} = 20\lg\left(\frac{1}{\sqrt{1 + \omega^2\tau^2}}\right) = -10\lg(1 + \omega^2\tau^2)$$

and a phase of $\tan^{-1} -\omega\tau$.

When $\omega \ll 1/\tau$ then $\omega^2\tau^2$ is negligible compared with 1 and so the gain is 0 dB. Thus, at low frequencies the Bode plot is a straight line with a constant value of 0 dB. For high frequencies when $\omega \gg 1/\tau$ then $\omega^2\tau^2$ is much greater than 1 and so we can neglect the 1 and the magnitude is $-20 \lg \omega\tau$. This is a straight line of slope -20 dB per decade which intersects with the zero decibel line when $\omega\tau = 1$. Figure 5.11 shows these lines for the low and high frequencies, their intersection being when $\omega = 1/\tau$; this intersection is termed the *break point* or *corner frequency*. The two lines are called the *asymptotic approximation* to the true plot. The true plot, when we do not make these approximations, differs slightly from the approximate plot and has a maximum error of 3 dB at the break point. Table 5.1 gives the errors in using the asymptotes.

At low frequencies, when ω is less than about $0.1/\tau$, the phase angle is virtually $0°$. At high frequencies, when ω is more than about $10/\tau$, it is virtually $-90°$. Between these frequencies the phase angle can be considered to give a reasonable straight line. The maximum error in assuming a straight line is $5\frac{1}{2}°$. When $\omega = 1/\tau$ then the phase angle is $45°$. Table 5.1 gives the errors in using the asymptotes.

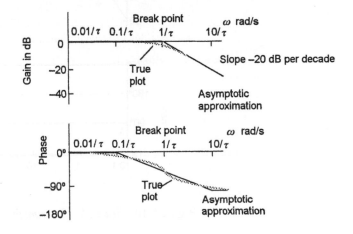

Figure 5.11 *Bode plot for transfer function $1/(1 + \tau s)$*

Table 5.1 *Asymptote errors for transfer function $1/(1 + \tau s)$*

ω	Magnitude error dB	Phase error
$0.10/\tau$	-0.04	$-5.7°$
$0.20/\tau$	-0.02	$+2.3°$
$0.50/\tau$	-1.0	$+4.9°$
$1.00/\tau$	-3.0	$0°$
$2.00/\tau$	-1.0	$-4.9°$
$5.00/\tau$	-0.2	$-2.3°$
$10.0/\tau$	-0.04	$+5.7°$

Example

Sketch the asymptotes of the Bode plots for a system having a transfer function of $100/(s + 100)$.

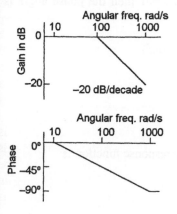

Figure 5.12 *Example*

The transfer function can be put in the form $1/(1 + s/100)$ and so the Bode plot is of the form shown in Figure 5.11. Since the time constant (1/100) then the break point is at $\omega = 100$ rad/s. At higher frequencies the slope of the gain asymptote will be -20 dB/decade and thus the gain plot is as shown in Figure 5.12. The phase is $\tan^{-1} -(\omega/100)$ and is thus $0°$ at low frequencies, $-45°$ at the break point and $90°$ at high frequencies; Figure 5.12 shows the plot.

5.4.5 Transfer function $(1 + \tau s)$

For $G(s) = (1 + \tau s)$, i.e. a zero at $s = 1/\tau$ where τ is the time constant, the frequency response is:

$$G(j\omega) = 1 + j\omega\tau$$

This means a gain in dB of:

$$\text{gain} = 20\lg\sqrt{1+\omega^2\tau^2} = 10\lg(1+\omega^2\tau^2)$$

and a phase of $\tan^{-1} \omega\tau$.

When $\omega \ll 1/\tau$ then $\omega^2\tau^2$ is negligible compared with 1 and so the gain is 0 dB. Thus at low frequencies the Bode plot is a straight line with a constant value of 0 dB. For high frequencies when $\omega \gg 1/\tau$ then $\omega^2\tau^2$ is much greater than 1 and so we can neglect the 1 and the magnitude is 20 lg $\omega\tau$. This is a straight line of slope $+20$ dB per decade which intersects with the zero decibel line when $\omega\tau = 1$. Figure 5.13 shows these lines, the *asymptotic approximation* to the true plot, for the low and high frequencies, their intersection being the *break point* or *corner frequency* of $\omega = 1/\tau$. The true plot has a maximum error which is 3 dB at the break point. The errors are the same as in Table 5.1.

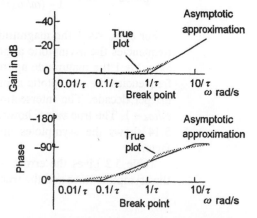

Figure 5.13 *Bode plot for transfer function $1/(1 + \tau s)$*

At low frequencies, when ω is less than about $0.1/\tau$, the phase angle is virtually $0°$. At high frequencies, when ω is more than about $10/\tau$, it is virtually $+90°$. Between these frequencies the phase angle can be considered to give a reasonable straight line. The maximum error in assuming a straight line is $5\frac{1}{2}°$. When $\omega\tau = 1$ then the phase angle is $45°$. Table 5.1 gives the errors.

5.4.6 Transfer function $\omega_n^2/(s^2 + 2\zeta\omega_n s + \omega_n^2)$

For a system having a transfer function:

$$G(s) = \frac{\omega_n^2}{s^2 + 2\zeta\omega_n s + \omega_n^2}$$

i.e. a pair of complex poles, the frequency response function is:

$$G(j\omega) = \frac{\omega_n^2}{-\omega^2 + j2\zeta\omega_n\omega + \omega_n^2}$$

$$= \frac{1}{[1 - (\omega/\omega_n)^2] + j[2\zeta(\omega/\omega_n)]}$$

$$= \frac{[1 - (\omega/\omega_n)^2] - j[2\zeta(\omega/\omega_n)]}{[1 - (\omega/\omega_n)^2]^2 + [2\zeta(\omega/\omega_n)]^2}$$

Thus the gain in decibels is:

$$\text{gain} = 20\lg\sqrt{\frac{1}{[1 - (\omega/\omega_n)^2]^2 + [2\zeta(\omega/\omega_n)]^2}}$$

$$= -10\lg\{[1 - (\omega/\omega_n)^2]^2 + [2\zeta(\omega/\omega_n)]^2\}$$

and the phase is:

$$\text{phase} = -\tan^{-1}\frac{2\zeta(\omega/\omega_n)}{1 - (\omega/\omega_n)^2}$$

For $\omega/\omega_n \ll 1$ the magnitude approximates to 0 dB and thus at low frequencies the asymptotic approximation is a straight line of 0 dB. For $\omega/\omega_n \gg 1$ the magnitude approximates to $-40\lg(\omega/\omega_n)$. Thus, at high frequencies the asymptotic approximation is a straight line of slope -40 dB per decade. The intersection of these two lines is a break point of $\omega/\omega_n = 1$. The true value, however, depends on the damping ratio. Figure 5.14 shows the asymptotes and some true plots at different damping factors.

Table 5.2 gives the errors, for a number of damping ratios, between the asymptote lines and the true magnitude plot.

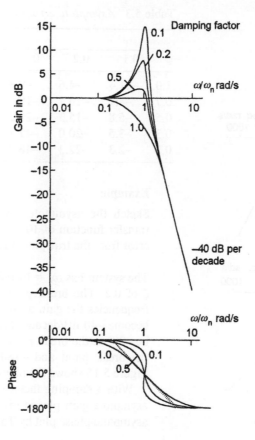

Figure 5.14   *Bode plot for a transfer function $\omega_n^2/(s^2 + 2\zeta\omega_n s + \omega_n^2)$*

Table 5.2   *Asymptote errors for gain in dB*

ζ	ω/ω_n						
	0.1	0.2	0.5	1.0	2.0	5.0	10.0
1.0	−0.09	−0.34	−1.94	−6.0	−1.92	−0.34	−0.09
0.7	0	−0.01	−0.26	−3.0	−0.26	−0.01	0
0.5	+0.04	+0.17	+0.90	0	+0.90	+0.17	+0.04
0.3	+0.07	+0.29	+1.85	+4.4	+1.85	+0.29	+0.07
0.2	+0.08	+0.33	+2.2	+8.0	+2.2	+0.33	+0.08

The phase is approximately constant at $0°$ for $\omega/\omega_n \ll 1$ and for $\omega/\omega_n \gg 1$ approximately $-180°$. Usually an asymptote line is drawn through the points $\omega/\omega_n = 0.2$ as $0°$ and $\omega/\omega_n = 5$ as $-180°$. The discrepancy between this line and the true phase plots is shown in Table 5.3.

Table 5.3   *Asymptote errors in degrees for phase*

ζ	0.1	0.2	0.5	1.0	2.0	5.0	10.0
1.0	−11.4	−4.6	−9.8	0	+9.8	+4.6	+11.4
0.7	−8.1	−10.7	−19.6	0	+19.6	+10.7	+ 8.1
0.5	−5.8	−15.3	−29.2	0	+29.2	+15.3	+5.8
0.3	−3.5	−20.0	−41.1	0	+41.1	+20.0	+3.5
0.2	−2.3	−22.3	−48.0	0	+48.0	+22.3	+2.3

(header spans ω/ω_n)

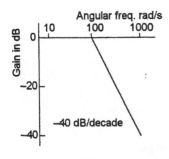

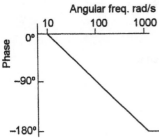

Figure 5.15   *Example*

Example

Sketch the asymptotes of the Bode plots for a system having a transfer function of $100/(s^2 + 4s + 100)$ and indicate the size of the error from the true plot at the break point.

The system has $\omega_n = 10$ rad/s and, since $2\zeta\omega_n = 4$, a damping factor ζ of 0.2. The break point is when $\omega = \omega_n = 10$ rad/s. For lower frequencies the gain asymptote will be 0 and at the break point will become −40 dB/decade. Figure 5.15 shows the plot. The phase will be −90° at the break point and effectively 0° one decade down from the break point and −180° one decade up from the break point. Figure 5.15 shows the plot.

With a damping factor of 0.2, the error at the break point for the asymptote gain plot is given by Table 5.2 as +8.0 dB and for the asymptote phase plot by Table 5.3 as zero.

5.4.6   Transfer function $(s^2 + 2\zeta\omega_n s + \omega_n{}^2)/\omega_n{}^2$

For a system having a transfer function:

$$G(s) = \frac{s^2 + 2\zeta\omega_n s + \omega_n^2}{\omega_n^2}$$

i.e. a pair of complex zeros, the frequency response is:

$$G(j\omega) = \frac{-\omega^2 + j2\zeta\omega_n\omega + \omega_n^2}{\omega_n^2}$$

$$= [1 - (\omega/\omega_n)^2] + j[2\zeta(\omega/\omega_n)]$$

The gain in decibels is thus:

$$\text{gain} = 20\lg\sqrt{[1 - (\omega/\omega_n)^2]^2 + [2\zeta(\omega/\omega_n)]^2}$$

$$= 10\lg\{[1 - (\omega/\omega_n)^2]^2 + [2\zeta(\omega/\omega_n)]^2\}$$

and the phase is:

$$\text{phase} = \tan^{-1}\left[\frac{2\zeta(\omega/\omega_\mathrm{n})}{1-(\omega/\omega_\mathrm{n})^2}\right]$$

The gain differs only from that in Section 5.4.5 in being positive rather than negative. Thus the magnitude plot is just the mirror image of Figure 5.14 about the 0 dB line. The phase differs from that in Section 5.4.5 in being positive rather than negative. Thus the phase plot is just the mirror image of Figure 5.14 about the 0° line. The differences of the true plots from the asymptote lines is the same as in Tables 5.2 and 5.3.

Example

Determine the asymptote Bode plot for the system having the transfer function:

$$G(s) = \frac{50(s+2)}{s(s+10)}$$

This can be considered to the multiplication of four elements:

$$G(s) = 10 \times (1 + \tfrac{1}{2}s) \times \frac{1}{s} \times \frac{1}{1+s/10}$$

We can draw the Bode plots for each of these elements and then sum them to obtain the overall plot. Figure 5.16 shows the result.

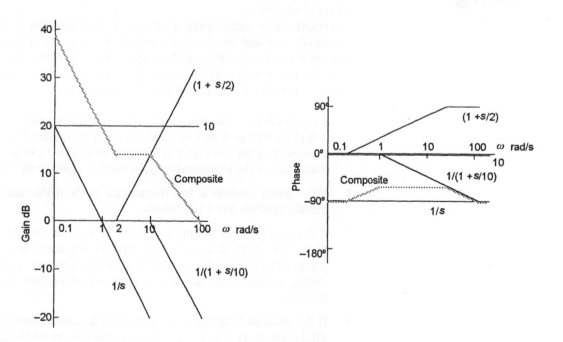

Figure 5.16 *Composite Bode plot*

1   For $G_1(s) = 10$ we have a straight line of magnitude $20 \lg 10 = 20$ dB and a constant phase of $0°$.

2   For $G_2(s) = 1 + \frac{1}{2}s$ we have a magnitude of 0 dB when $\omega\tau \ll 1$ and a line of slope $+20 \lg \omega\tau = +20$ dB per decade when $\omega\tau \gg 1$. The break point is $\omega = 1/\tau = 2$ rad/s. The phase is effectively $0°$ up to $0.1/\tau = 0.2$ rad/s and $+90°$ for frequencies greater than $10/\tau = 20$ rad/s.

3   For $G_3(s) = 1/s$ we have a straight line of slope -20 dB per decade passing through the 0 dB point at $\omega = 1$ rad/s. The phase is a constant $-90°$.

4   For $G_4 = 1/(1 + s/10)$ we have a magnitude of 0 dB when $\omega\tau \ll 1$ and a line of slope $-20 \lg \omega\tau = -20$ dB/decade when $\omega\tau \gg 1$. The break point is $\omega = 1/\tau = 10$ rad/s. The phase is effectively $0°$ up to $0.1/\tau = 1$ rad/s and $-90°$ for frequencies greater than $10/\tau = 100$ rad/s.

5.5 System identification

In Chapter 2, methods were indicated by which models, i.e. differential equations describing the input–output relationship or transfer function, can be devised for systems by considering them to be made up of simple elements. An alternative way of developing a model for a real system is determine its response to some input and then find the model that fits the response; this process of determining a mathematical model is known as *system identification*.

A particularly useful method of system identification is to use a sinusoidal input and determine the output over a range of frequencies. Bode plots are then plotted with this experimental data. We then find the Bode plot elements that fit the experimental plot by drawing asymptotes on the gain Bode plot and considering their gradients.

1   If the gradient at low frequencies prior to the first corner frequency is zero then there is no s or $1/s$ element in the transfer function. The K element in the numerator of the transfer function can be obtained from value of the low frequency gain since the gain in dB = $20 \lg K$.

2   If the initial gradient at low frequencies is -20 dB/decade then the transfer function has a $1/s$ element.

3   If the gradient becomes more negative at a corner frequency by 20 dB/decade, there is a $(1 + s/\omega_c)$ term in the denominator of the transfer function, with ω_c being the corner frequency at which the change occurs. Such terms can occur for more than one corner frequency.

4   If the gradient becomes more positive at a corner frequency by 20 dB/decade, there is a $(1 + s/\omega_c)$ term in the numerator of the transfer function, with ω_c being the frequency at which the change occurs. Such terms can occur for more than one corner frequency.

5   If the gradient at a corner frequency becomes more negative by 40 dB/decade, there is a $(s^2/\omega_c^2 + 2\zeta s/\omega_c + 1)$ term in the denominator of the transfer function. The damping ratio ζ can be found from considering the behaviour of the system to a unit step input.

6   If the gradient at a corner frequency becomes more positive by 40 dB/decade, there is a $(s^2/\omega_c^2 + 2\zeta s/\omega_c + 1)$ term in the numerator of the transfer function. The damping ratio ζ can be found from considering the detail of the Bode plot at a corner frequency.

7   If the low-frequency gradient is not zero, the K term in the numerator of the transfer function can be determined by considering the value of the low-frequency asymptote. At low frequencies, many terms in transfer functions can be neglected and the gain in dB approximates to 20 lg (K/ω^2). Thus, at $\omega = 1$ the gain in dB approximates to 20 lg K.

The phase angle curve is used to check the results obtained from the magnitude analysis.

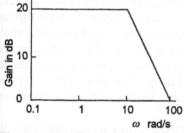

Example

Determine the transfer function of the system giving the Bode magnitude plot shown in Figure 5.17.

The initial gradient is 0 and so there is no $1/s$ or s term in the transfer function. The initial gain is 20 and thus $20 = 20$ lg K and so we have $K = 10$. The gradient changes by -20 dB/decade at a frequency of 10 rad/s. Hence there is a $(1 + s/10)$ term in the denominator. The transfer function is thus $10/(1 + 0.1s)$.

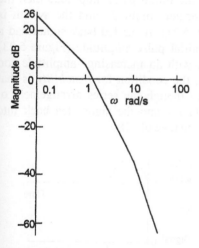

Example

Determine the transfer function of the system giving the Bode magnitude plot shown in Figure 5.18.

There is an initial slope of -20 dB/decade and so a $1/s$ term. At the corner frequency 1.0 rad/s there is a -20 dB/decade change in gradient and so a $1/(1 + s/1)$ term. At the corner frequency 10 rad/s there is a further -20 dB/decade change in gradient and so a $1/(1 + s/10)$ term. At $\omega = 1$ the magnitude is 6 dB and so $6 = 20$ lg K and $K = 10^{6/20} = 2.0$. The transfer function is thus $2.0/s(1 + s)(1 + 0.1s)$.

Example

Determine the transfer function of the system giving the Bode magnitude plot shown in Figure 5.19. This shows both the asymptotes and the departure of the true plot from them in the vicinity of the break point.

The gain Bode plot has an initial zero gradient. Since the initial magnitude is 10 dB then $10 = 20$ lg K and so $K = 10^{0.5} = 3.2$. The

Figure 5.17   *Example*

Figure 5.18   *Example*

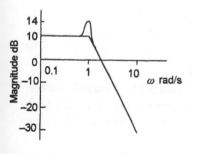

Figure 5.19   *Example*

change of –40 dB/decade at 1 rad/s means there is a $1/(s^2 + 2\zeta s + 1)$ term. The transfer function is thus $3.2/(s^2 + 2\zeta s + 1)$.

The damping factor ζ can be obtained by considering the departure of the true Bode plot from the asymptotes at the break point. Since it rises by about 4 dB, Table 5.2 indicates that this corresponds to a damping factor of about 0.3. The transfer function is thus $3.2/(s^2 + 0.6s + 1)$.

5.6 Stability

Consider the stability of a closed-loop control system (Figure 5.20) when we have a brief input of a pulse. We will regard the pulse as essentially half of a sinusoidal signal at a particular frequency (Figure 5.21(a)). This passes through $G(s)$ to give an output which is then fed back through $H(s)$. Suppose it arrives back with amplitude unchanged from that of the input but with a phase such that when it is subtracted from the now zero input signal we have a resulting error signal which just continues the initial half-rectified pulse (Figure 5.21(b)). This then continues round the feedback loop to once again arrive just in time to continue the signal. There is a self-sustaining oscillation.

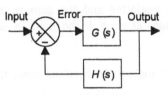

Figure 5.20   *Closed-loop system*

If the fed back signal has an amplitude which was smaller than the initial pulse amplitude then the signal would die away with time (Figure 5.22). The system is *stable*. If the fed back signal had an amplitude which was just the same as that of the initial pulse amplitude then the oscillation would continue with constant amplitude and the system is said to be *marginally stable* (Figure 5.21). If the fed back signal had a larger amplitude than that of the initial pulse amplitude (Figure 5.23) then the oscillation would continue with an increasing amplitude and the system would be unstable. Thus the condition for *instability* is that the gain resulting from a signal fed though the series arrangement of $G(s)$ and $H(s)$ should be greater than 1 and the signal fed back into $G(s)H(s)$ must have suffered a phase change of –180°.

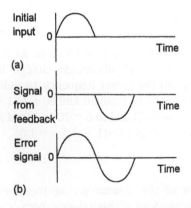

Figure 5.21   *Self sustaining signal: (a) input to system, (b) fed back signal, (c) resulting error signal*

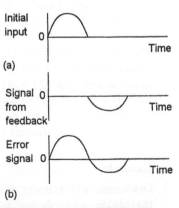

Figure 5.22   *Stable system: (a) input to system (b) fed back, signal, (c) resulting error signal*

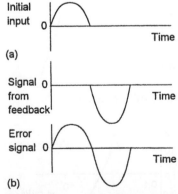

Figure 5.23   *Unstable system: (a) input to system (b) fed back, signal, (c) resulting error signal*

The transfer function for $G(s)$ in series with $H(s)$ is called the *open-loop transfer function*. The stability criterion can thus be stated as:

The critical point which separates stable from unstable systems is when the open-loop phase shift is −180° and the open-loop gain is 1.

A good stable control system usually has an open-loop gain significantly less than 1, typically about 0.4 to 0.5, when the phase shift is −180° and an open-loop phase shift of between −115° to −125° when the gain is 1. Such values give a slightly under damped system which gives, with a step input, about a 20 to 30% overshoot.

5.6.1 Stability measures

Measures of the stability of systems that are used in the frequency domain are:

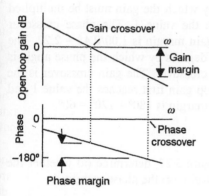

Figure 5.25   *Stability and Bode plots*

1   *Phase crossover frequency*
    The phase crossover frequency is the frequency at which the phase angle first reaches −180°.

2   *Gain margin*
    This is the factor by which the gain must be multiplied at the phase crossover to have the value 1. A good stable control system usually has an open-loop gain significantly less than 1, typically about 0.4 to 0.5, when the phase shift is −180° and so a gain margin of 1/0.5 to 1/0.4, i.e. 2 to 2.5.

3   *Gain crossover*
    This is the frequency at which the open-loop gain first reaches the value 1.

4   *Phase margin*
    This is the number of degrees by which the phase angle is smaller than −180° at the gain crossover. A good stable control system usually has typically an open-loop phase shift of between −115° to −125° when the gain is 1; thus, the phase margin is between 45° and 65°.

The above parameters enable the questions of how much change in gain and phase of the $G(s)H(s)$ product can be tolerated before a system becomes unstable. They are thus useful in the design of stable systems.

The Bode plot for the open-loop transfer function, i.e. $G(s)H(s)$, gives a convenient way to determine the above parameters and hence the stability of a system. An open-loop gain of 1 is, on the log scale of dB, a gain of 20 lg 1 = 0 dB. Figure 5.25 shows the parameters on a Bode plot. Note that on the Bode gain plot we are working with the log of the gain and so the gain margin is the additional dB that is necessary to make the gain of the signal unity at the phase crossover frequency.

Figure 5.26(a) is an example of Bode plot for a stable system with Figure 5.26(b) being for an unstable system. In (a), the open-loop gain is less than 0 dB, i.e. a gain of 1, when the phase is −180°. In (b), the

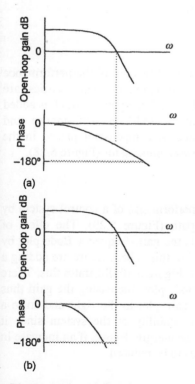

Figure 5.26   *(a) Stable system, (b) unstable system*

open-loop gain is greater than 0 dB, i.e. greater than 1, when the phase is −180°.

Example

Determine the gain margin and the phase margin for a system that gave the following open-loop experimental frequency response data: at frequency 0.005 Hz a gain of 1.00 and phase −120°, at 0.010 Hz a gain of 0.45 and phase −180°.

The gain margin is the factor by which the gain must be multiplied at the phase crossover to have the value 1. The phase crossover occurs at 0.010 Hz and so the gain margin is 1.00/0.45 = 2.22. The phase margin is the number of degrees by which the phase angle is smaller than −180° at the gain crossover. The gain crossover is the frequency at which the open-loop gain first reaches the value 1 and so is 0.005 Hz. Thus, the phase margin is 180° − 120° = 60°.

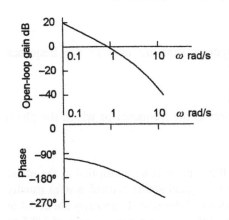

Figure 5.27   *Example*

Example

For the Bode plot shown in Figure 5.27, determine (a) whether the system is stable, (b) the gain margin, (c) the phase margin.

(a) The system is stable because it has an open-loop gain less than 1 when the phase is −180°.
(b) The gain margin is about 12 dB.
(c) The phase margin is about 30°.

5.7 Compensation

The term *compensation* is used for the modification of the performance characteristics of a system so that the required characteristics are obtained. A *compensator* is thus an additional component which is added into a control system to modify the closed-loop performance and compensate for a deficient performance. A compensator placed in the forward path is called a *cascade* or *series* compensator (Figure 5.28).

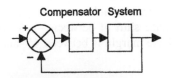

Figure 5.28   *Cascade compensation*

5.7.1 Changing the gain

Consider the effects of adjusting the performance of a control system by changing the gain in the forward path (Figure 5.29). The effect of increasing the gain is to shift upwards the gain–frequency Bode plot by an equal amount over all the frequencies; this is because we are adding a constant gain element to the Bode plot. Figure 5.30 illustrates this. There is no effect on the phase–frequency Bode plot. Increasing the gain thus shifts the 0 dB crossing point of the gain plot to the right and so to a higher frequency. This decreases the stability of the system since it decreases the gain margin and the phase margin. Hence, if an increase in stability is required then the gain needs to be reduced.

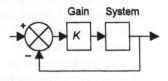

Figure 5.29   *Adjusting the performance of a system by introducing gain*

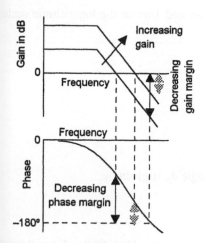

Figure 5.30 *The effect of increasing the gain*

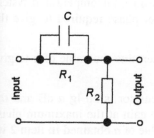

Figure 5.32 *Phase-lead compensator*

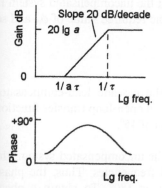

Figure 5.33 *Bode plot for a phase-lead compensator*

Example

For the control system giving the open-loop Bode plot of Figure 5.31 we have a control system with a phase margin of 35°. By how much should the gain of the system be changed if a gain margin of 45° is required?

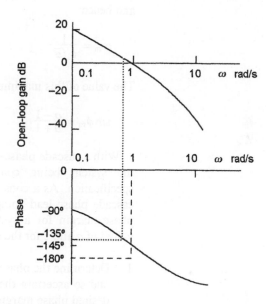

Figure 5.31 *Example*

The phase margin of 35° means that the phase is 180° – 35° = 145° when the gain is 0 dB. For a phase margin of 45° the phase must be 180° – 45° = 135° when the gain is 0 dB. At present, when the phase is 135° the gain is about +2 dB. If follows that if the gain is reduced by 2 dB that the gain–frequency line will be shifted downwards by 2 dB and give the required phase margin. Since the change in gain ΔK is given by $2 = 20 \lg \Delta K$ then $\Delta K = 1.3$. This is the factor by which the gain has to be reduced.

5.7.2 Phase-lead compensation

The transfer function of a phase-lead compensator is of the form:

$$G(s) = \frac{1 + a\tau s}{1 + \tau s}$$

with $a > 1$. Such a compensator can be provided by the circuit shown in Figure 5.32 ($a = (R_1 + R_2)/R_2$ and $\tau = R_1 R_2 C/(R_1 + R_2)$). Figure 5.33 shows the Bode plot for a phase-lead compensator (it can be obtained by adding the Bode plots for the numerator term and the denominator term). The term *phase-lead* is used because the compensator has a positive phase and so is used to add phase to an uncompensated system. The maximum value of the phase occurs at a frequency ω_m which is midway between the frequencies of $1/\tau$ and $1/a\tau$ on the logarithmic scale and so:

$$\lg \omega_m = \frac{1}{2}\left(\lg \frac{1}{\tau} + \lg \frac{1}{a\tau}\right)$$

and hence:

$$\omega_m = \frac{1}{\tau\sqrt{a}}$$

The value of this maximum phase angle ϕ_m is given by:

$$\sin \phi_m = \frac{a-1}{a+1}$$

With a cascade phase-lead compensator we add its Bode plot to that of the system being compensated in order to obtain the required specification. As a consequence we can increase the phase margin. The cascade phase-lead compensator is thus used to provide a satisfactory phase margin for a system. The procedure to determine the required values of a and τ for the compensator is to:

1   Determine the phase margin of the open-loop uncompensated system and so ascertain the addition amount of phase required to give the desired phase margin.

2   The above equation is then used to determine the value of a to give this additional phase.

3   The high frequency gain of the compensator is 20 lg a dB and the low frequency gain is 0 dB; hence, the gain at the maximum phase is ½ 20 lg a = 10 lg a dB. From the value of a obtained in item 2 we can determine this gain.

4   The compensator is used to give the new gain crossover at the frequency at which the phase is a maximum and thus we need to place this frequency where the gain of the uncompensated system is -10 lg a dB. In doing this we obtain the required value of ϕ_m and so can determine the required value of τ.

Example

Determine the transfer function of the cascade lead-compensator that can be used with a system having an open-loop transfer function of $10/s^2$ in order to give a phase margin of 45°.

Figure 5.34 shows the Bode plot for the uncompensated system, the uncompensated phase being 0° at all frequencies. Thus, the phase margin of the uncompensated system is 180°. To obtain a phase margin of 45° we require:

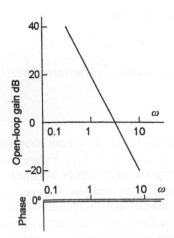

Figure 5.34   *Example*

$$\sin 45^\circ = \frac{a-1}{a+1}$$

$$0.71(a+1) = a-1$$

$$0.29a = 1.71$$

and thus a is 5.9. The gain of the compensator at the maximum compensator phase is $10 \lg a = 7.7$ dB. The uncompensated system has a gain of -7.7 dB at a frequency of about 5 rad/s and so $\omega_m = 5 = 1/\tau\sqrt{a} = 1/2.4\tau$ and thus $\tau = 0.083$. The required compensator has thus a transfer function of:

$$G(s) = \frac{1+a\tau s}{1+\tau s} = \frac{1+0.49s}{1+0.083s}$$

5.7.3 Phase-lag compensation

The phase-lag compensator has a negative phase angle and so is used to subtract phase from an uncompensated system. A phase-lag compensator has a transfer function of the form:

$$G(s) = \frac{1+\tau s}{1+a\tau s}$$

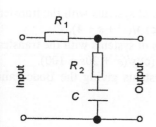

where a is greater than 1. Figure 5.35 shows the type of circuit that can be used ($a = (R_1 + R_2)/R_2$ and $\tau = R_2C$). Figure 5.36 shows the Bode plot. Because the phase-lag compensator adds a negative phase angle to a system, the phase lag is not a useful effect of the compensation and does not provide a direct means of improving the phase margin. The phase-lag compensator does, however, reduce the gain and so can be used to lower the crossover frequency. A consequence of this is that, as usually the phase margin of the system is higher at the lower frequency, the stability can be improved. The procedure that can be adopted to design with a cascade phase-lag compensator is:

Figure 5.35 *Phase-lag compensator*

1   Determine the frequency where the required phase margin would be obtained if the gain plot crossed the 0 dB line at this frequency. Allow for 5° phase lag from the phase-lag compensator when determining the new crossover frequency.

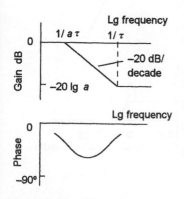

Figure 5.36 *Phase-lag compensator*

2   One decade below the new crossover frequency is taken to be $1/\tau$. It is necessary to ensure that the phase minimum occurs at a frequency which is well below the crossover frequency of the uncompensated system so that the phase lag introduced by the compensator does not significantly destabilise the system.

3   Determine the gain required at the new crossover frequency to ensure that the compensated system gain plot crosses at this frequency. Since the gain produced by the phase-lag compensator at this frequency is $-20 \lg a$, we can calculate a.

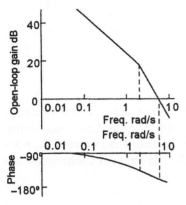

Figure 5.37 *Example*

Example

Determine the transfer function of the cascade phase-lag compensator that can be used with a system having an open-loop Bode plot of Figure 5.37 in order to give a phase margin of 40°.

The uncompensated system has a phase margin of about 20°. The frequency where the phase margin is the required 40° + 5° is 2 rad/s and this is to become the new crossover frequency. One decade below this is 0.2 rad/s. Thus $1/\tau = 0.2$ and so $\tau = 5$. The gain of the uncompensated system at the new crossover frequency is 20 dB and so 20 lg a = 20 and hence a = 10. Thus, the required transfer function of the compensator is $(1 + 5s)/(1 + 50s)$.

Problems

1 What are the magnitudes and phases of the signals represented by phasors described as (a) j2, (b) $j^2 2$, (c) 2 + j1?

2 What are the frequency response functions for systems with transfer functions (a) $1/(s + 5)$, (b) $7/(s + 2)$, (c) $1/[(s + 10)(s + 2)]$?

3 Determine the magnitude and the phase of the response of a system with transfer function $3/(s + 2)$ to sinusoidal inputs of angular frequency (a) 1 rad/s, (b) 2 rad/s.

4 Sketch the asymptotes for the Bode plots of systems with the transfer functions (a) 100, (b) $1000/(s + 1000)$, (c) $4/(s^2 + s + 4)$.

5 Sketch the asymptotes for the Bode plots of systems with the transfer function (a) $10/s^2$, (b) $(s - 10)/(s + 10)$, (c) $s/(s^2 + 20s + 100)$.

6 Obtain the transfer functions of the systems giving the Bode gain plots in Figure 5.38.

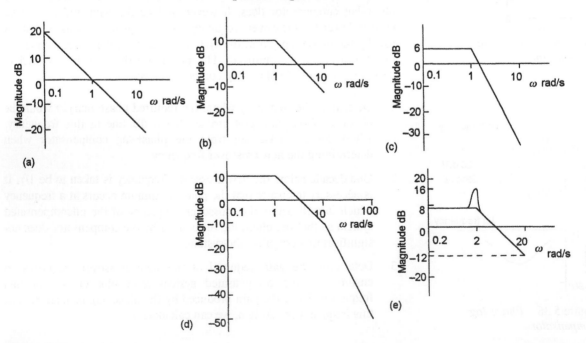

Figure 5.38 *Problem 6*

7 The following are experimentally determined frequency response data for a system. By plotting the Bode gain diagram, determine the transfer function of the system.

Freq. Hz	0.16	0.47	1.3	2.5	4.8	10.0	16.0	20.0	24.0
Gain dB	24.0	24.0	23.6	23.0	17.0	14.0	9.5	8.0	5.0

8 Determine the gain margin and the phase margin for a system that gave the following open-loop experimental frequency response data: at frequency 0.01 Hz a gain of 1.00 and phase −130°, at 0.02 Hz a gain of 0.55 and phase −180°.

9 For the Bode plot shown in Figure 5.39, determine (a) whether the system is stable, (b) the gain margin, (c) the phase margin.

10 A system has an open-loop transfer function of $1/[s(1 + s)(1 + 0.2s)]$. By plotting the Bode diagrams, determine (a) whether the system is stable, (b) the gain margin, (c) the phase margin.

11 A system has an open-loop transfer function of $1/[s(1 + 0/0.02s)(1 + 0.2s)]$. By plotting the Bode diagrams, determine (a) whether the system is stable, (b) the gain margin, (c) the phase margin.

12 For the control system giving the open-loop Bode plot of Figure 5.40 we have a control system with a phase margin of 10°. By how much should the gain of the system be changed if a gain margin of 40° is required?

13 The following are experimentally determined open-loop frequency response data for a system. Determine the phase margin when uncompensated and the change in the gain of the system that is necessary from a phase-lead compensator, when added in cascade, to increase the phase margin by 5°.

Frequency rad/s	0.6	0.8	1.0	2.0	4.0	6.0
Gain dB	13.2	10.3	8.0	0.0	−10.7	−18.0
Phase deg.	−110	−116	−122	−146	−175	−192

14 Determine the maximum phase lead introduced by a phase-lead compensator with a transfer function of $(1 + 0.2s)/(1 + 0.02s)$.

15 A system has an open-loop transfer function of $12/[s(s + 1)]$. What will be the transfer function of a phase-lead compensator which, when added in cascade, will increase the uncompensated phase margin of 15° to 40°?

16 A system has an open-loop transfer function of $4/[s(2s + 1)]$. What will be the transfer function of a phase-lag compensator which, when added in cascade, will give a phase margin of 40°?

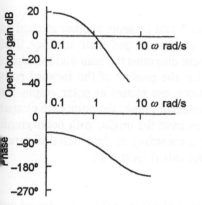

Figure 5.39   *Problem 9*

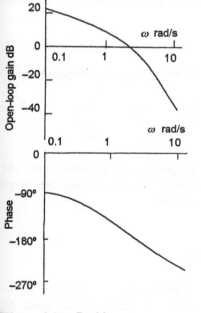

Figure 5.40   *Problem 12*

6 Nyquist diagrams

6.1 Introduction

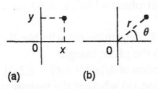

(a)   (b)

Figure 6.1 *(a) Cartesian graph with points specified by x and y values, (b) polar graph with points specified by r and θ values*

This chapter follows on from Chapter 5 and presents another method of considering the frequency response of systems and their stability. The method uses *Nyquist diagrams*; in these diagrams the gain and the phase of the open-loop transfer function, i.e. the product of the forward path and the feedback path transfer functions, are plotted as polar graphs for various values of frequency. With Cartesian graphs the points are plotted according to their x and y coordinates from the origin; with polar graph the points are plotted from the origin according to their radial distance from it and their angle to the reference axis (Figure 6.1).

6.2 The polar plot

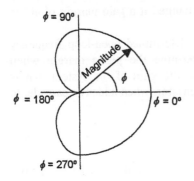

Figure 6.2 *Polar plot with the plot as the line traced out by the tips of the phasors as the frequency is changed from zero to inifnity*

The *polar plot* of the frequency response of a system is the line traced out as the frequency is changed from 0 to infinity by the tips of the phasors whose lengths represent the magnitude, i.e. amplitude gain, of the system and which are drawn at angles corresponding to their phase (Figure 6.2).

Example

Draw the polar diagram for the following frequency response data.

Freq. rad/s	1.4	2.0	2.6	3.2	3.8
Magnitude	1.6	1.0	0.6	0.4	0.2
Phase deg.	−150	−160	−170	−180	−190

Note that the above data gives the phases with negative signs. This means they are lagging behind the 0° line by the amounts given Figure 6.3 shows the polar plot obtained.

6.2.1 Nyquist diagrams

The term *Nyquist diagram* is used for:

> The Nyquist diagram is the line joining the series of points plotted on a polar graph when each point represents the magnitude and phase of the open-loop frequency response corresponding to particular frequency.

To plot the Nyquist diagram from the open-loop transfer function of system we need to determine the magnitude and the phase as functions of frequency.

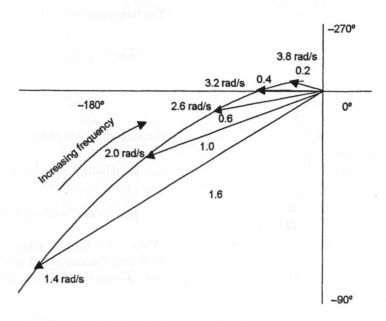

Figure 6.3 *Example*

Example

Determine the Nyquist diagram for a first-order system with an open-loop transfer function of $1/(1+\tau s)$.

The frequency response is:

$$\frac{1}{1+j\omega\tau} = \frac{1}{1+j\omega\tau}\frac{1-j\omega\tau}{1-j\omega\tau} = \frac{1}{1+\omega^2\tau^2} - j\frac{\omega\tau}{1+\omega^2\tau^2}$$

The magnitude is thus:

$$\text{magnitude} = \frac{1}{\sqrt{1+\omega^2\tau^2}}$$

and the phase is:

$$\text{phase} = -\tan^{-1}\omega\tau$$

At zero frequency the magnitude is 1 and the phase 0°. At infinite frequency the magnitude is zero and the phase is −90°. When $\omega\tau = 1$ the magnitude is $1/\sqrt{2}$ and the phase is −45°. Substitution of other values leads to the result shown in Figure 6.4 of a semicircular plot.

Example

Determine the Nyquist plot for the system having the open-loop transfer function of $1/s(s + 1)$.

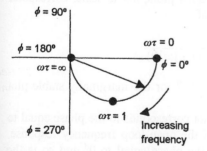

Figure 6.4 *Nyquist diagram for a first-order system*

The frequency response is:

$$G(j\omega) = \frac{1}{j\omega(j\omega+1)} = \frac{1}{j\omega - \omega^2} = \frac{1}{j\omega - \omega^2}\frac{-j\omega - \omega^2}{-j\omega - \omega^2}$$

$$= \frac{-\omega^2}{\omega^2 + \omega^4} - j\frac{\omega}{\omega^2 + \omega^4}$$

the magnitude and phase are thus:

$$\text{magnitude} = \frac{1}{\omega\sqrt{\omega^2 + 1}}$$

$$\text{phase} = \tan^{-1} -1/(-\omega) = 180° + \tan^{-1}1/\omega$$

When $\omega = \infty$ then the magnitude is 0 and the phase is 0°. As ω tends to 0 then the magnitude tends to infinity and the phase to 270° or –90°. Figure 6.5 shows the polar plot.

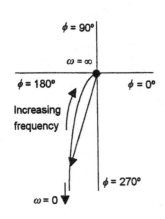

Figure 6.5   *Example*

6.3 Stability

As indicated in Section 5.6: the critical point which separates stable from unstable systems is when the open-loop phase shift is –180° and the open-loop magnitude is 1. If a Nyquist diagram of the open-loop frequency response is plotted then for the system to be stable there must not be any phasor with length greater than 1 and phase –180°. Thus the line traced by the tips of the phasors, the so-termed loci, must not enclose the –1 point.

> *Closed-loop systems whose open-loop frequency response $G(j\omega)H(j\omega)$ loci, as ω goes from 0 to ∞, do not encircle the –1 point will be stable, those which encircle the –1 point are unstable and those which pass through the –1 point are marginally stable. Encircling the point may be taken as passing to the left of the point.*

The above statement is known as the *Nyquist stability criterion*.

Figure 6.6 illustrates the above with examples of stable, marginally stable and unstable systems. The Nyquist plots, not to scale, correspond to the open-loop frequency response of:

$$G(j\omega)H(j\omega) = \frac{K}{(1 + j\omega0.2)(1 + j\omega)(1 + j\omega10)}$$

with $K = 10$ for the stable plot, $K = 137$ for the marginally stable plot and $K = 500$ for the unstable plot.

The vertical axis of the Nyquist plot corresponds to the phase equal to 90° and so is the imaginary part of the open-loop frequency response. The horizontal axis corresponds to the phase equal to 0° and so is the real part of the open-loop frequency response.

Figures 6.7 to 6.10 shows examples of Nyquist plots for common forms of open-loop transfer functions and their conditions for stability.

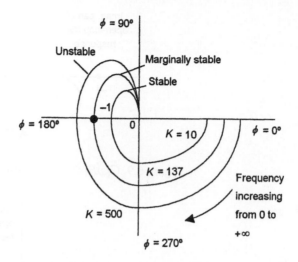

Figure 6.6 *Stability and the Nyquist plot*

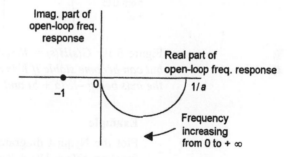

Figure 6.7 $G(s)H(s) = K/(s + a)$, *stable for all values of K > 0*

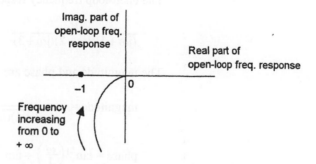

Figure 6.8 $G(s)G(s) = K/s(s + a)$, *stable for all values of K > 0*

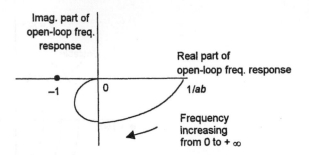

Figure 6.9 *G(s)H(s) = K/(s + a)(s + b), stable for all values of K > 0*

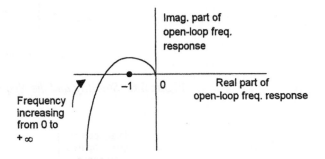

Figure 6.10 *G(s)H(s) = K/s(s + a)(s + b); this is unstable with large K but can become stable if K is reduced, the point at which the plot crosses the axis being –K/(a + b) and so stability is when –K/(a + b) > –1*

Example

Plot the Nyquist diagram for a system with the open-loop transfer function $K/(s + 1)(s + 2)(s + 3)$ and consider the value of K needed for stability.

The open-loop frequency response is:

$$\frac{K}{(j\omega + 1)(j\omega + 2)(j\omega + 3)}$$

The magnitude and phase are:

$$\text{magnitude} = \frac{K}{\sqrt{(\omega^2 + 1)(\omega^2 + 4)(\omega^2 + 9)}}$$

$$\text{phase} = \tan^{-1}\left(\frac{\omega}{1}\right) + \tan^{-1}\left(\frac{\omega}{2}\right) + \tan^{-1}\left(\frac{\omega}{3}\right)$$

When $\omega = 0$ then the magnitude is $K/6$ and the phase is 0°. When $\omega = \infty$ then the magnitude is 0 and the phase is 270°. We can use these, and other points to plot the polar graph.

Alternatively we can consider the frequency response in terms of real and imaginary parts. We can write the open-loop frequency function as:

$$\frac{6K(1-\omega^2)}{(\omega^2+1)(\omega^2+4)(\omega^2+9)} + j\frac{\omega K(\omega^2-11)}{(\omega^2+1)(\omega^2+4)(\omega^2+9)}$$

When $\omega = 0$ then the imaginary part is zero and the real part is $K/6$. When $\omega = \infty$ then the imaginary part is zero and the real part is 0. The imaginary part will be zero when $\omega = \sqrt{11}$. This is a real part, and hence magnitude, of $-K/60$ and is the point at which the plot crosses the real axis. Thus for a stable system we must have $-K/60$ less than -1, i.e. K must be less than 60. Figure 6.11 shows the complete Nyquist plot (not to scale).

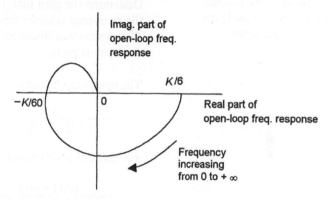

Figure 6.11   *Example*

6.4 Relative stability

The use of gain margin and phase margin was introduced in Section 5.6.1 to discuss the relative stability of a system in the frequency domain when described by a Bode plot. With Nyquist plots:

1   *Gain margin*

The phase crossover frequency is the frequency at which the phase angle first reaches $-180°$ and thus is the point where the Nyquist plot crosses the real axis (Figure 6.12). On a Nyquist plot the $(-1, j0)$ point is the point separating stability from instability. The gain margin is the amount by which the actual gain must be multiplied before the onset of instability. Thus if the plot cuts the negative real axis at $-x$ (Figure 6.12), it has to be multiplied by $1/x$ to give the value -1 and so the gain margin, which is expressed in dB, is $20 \lg(1/x)$.

When the open-loop plot goes through the $(-1, j0)$ point the gain margin is 0 dB, the system being on the margin of instability. When the open-loop plot goes to the left of $(-1, j0)$ point the gain margin is negative in dB, the system being unstable. When the open-loop plot goes to the right of $(-1, j0)$ point the gain margin is positive in dB,

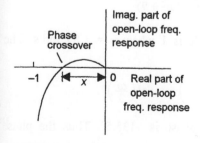

Figure 6.12   *Phase crossover and gain margin*

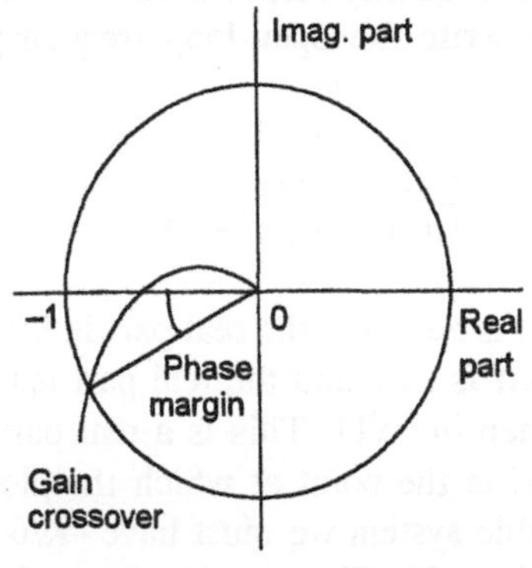

Figure 6.13   *Phase margin: the angle through which the gain crossover line must be rotated to reach the real axis and pass through the (–1, j0) point*

the system being stable. When the open-loop plot does not intersect the negative real axis the gain margin is infinite in dB.

2   *Phase margin*

The phase margin is defined as the angle in degrees by which the phase angle is smaller than –180° at the gain crossover, the gain crossover being the frequency at which the open-loop gain first reaches 1. Thus, with a Nyquist plot, if we draw a circle of radius 1 centred on the origin, then the point at which it intersects the Nyquist line gives the gain crossover. The phase margin is the angle through which this gain crossover line must be rotated about the origin to reach the real axis and pass through the (–1, j0) point (Figure 6.13).

Example

Determine the gain margin and the phase margin for a system with the open-loop transfer function $K/(s + 1)(s + 2)(s + 3)$ with $K = 20$. This system was discussed earlier in this chapter (see Figure 6.11 for the Nyquist plot).

The open-loop frequency response is:

$$\frac{K}{(j\omega + 1)(j\omega + 2)(j\omega + 3)}$$

and this can be rearranged to give:

$$\frac{6K(1 - \omega^2)}{(\omega^2 + 1)(\omega^2 + 4)(\omega^2 + 9)} + j\frac{\omega K(\omega^2 - 11)}{(\omega^2 + 1)(\omega^2 + 4)(\omega^2 + 9)}$$

The imaginary part will be zero when $\omega = \sqrt{11}$ and thus the real part is $-K/60$ and is the point at which the plot crosses the real axis. Hence, if we have $K = 20$ then the plot intersects the negative real axis at $-20/60 = -1/3$. The gain can thus be increased by a factor of 3 in order to reach the –1 point. The gain margin is thus 20 lg 3 = 9.5 dB.

The magnitude is:

$$\text{magnitude} = \frac{K}{\sqrt{(\omega^2 + 1)(\omega^2 + 4)(\omega^2 + 9)}}$$

Thus, for $K = 20$, the magnitude is 1 when $\omega = 1.84$ rad/s. The phase is given by:

$$\text{phase} = \tan^{-1}\left(\frac{\omega}{1}\right) + \tan^{-1}\left(\frac{\omega}{2}\right) + \tan^{-1}\left(\frac{\omega}{3}\right)$$

and so, at this frequency, the phase is –135.5°. Thus the phase margin is 44.5°.

Problems

1 Sketch the Nyquist diagram for a system having an open-loop transfer function of $1/[s(s + 1)]$.

2 With a Nyquist diagram for the open-loop frequency response for a system, what is the condition for the system to be stable?

3 Determine the gain margin and the phase margin for a system which gave the following open-loop frequency response:

Freq. rad/s	1.4	2.0	2.6	3.2	3.8
Magnitude	1.6	1.0	0.6	0.4	0.2
Phase deg.	−150	−160	−170	−180	−190

4 Determine the gain margin and the phase margin for a system which gave the following open-loop frequency response:

Freq. rad/s	4	5	6	8	10
Gain	3.2	2.3	1.7	1.0	0.6
Phase in deg.	−140	−150	−157	−170	−180

5 Determine the gain margin and the phase margin for a system having an open-loop transfer function of:

$$\frac{K}{s(s + 1)(s + 2)}$$

when $K = 4$.

6 Determine the gain margin and the phase margin for a system having an open-loop transfer function of :

$$\frac{1}{s(0.2s + 1)(0.05s + 1)}$$

7 Controllers

7.1 Introduction

Process controllers are components which basically have an input of the error signal, i.e. the difference between the required value signal and the feedback signal, and an output of a signal to modify the system output. The simplest form of controller is an on–off device which switches on some correcting device when there is an error and switches it off when the error ceases. However, such a method of control has limitations and often more sophisticated controllers are used. While there are many ways a controller could be designed to react to an error signal, a form of controller which can give satisfactory control in a wide number of situations is the *three-term* or *PID controller*. The term *control mode* is used for the type of response a controller gives to an error signal and the three basic modes that are used are proportional (P), integral (I) and derivative (D); the three-term controller is a combination of all three modes.

The chapter is a discussion of process controllers and the modes of control law used, also including a brief consideration of programmable logic controllers and embedded microprocessor-based controllers.

7.1.1 Compensation and controllers

The term *compensation* is used for the modification or compensation of the performance characteristics of a system so that the required characteristics are obtained. *Compensators* are components which are added to a control system in order to modify closed-loop performance. They can be added anywhere in a control system. *Controllers* are components which basically have an input of the error signal and an output of a signal to modify the system output and give the required characteristics. They are thus used at a specific point in a control circuit. From the design point of view there is no real difference between compensators and controllers, the two terms reflecting differences in hardware. Traditionally a controller is a stand-alone component offering a range of control modes such as proportional gain and integral and derivative action and normally includes a summing element.

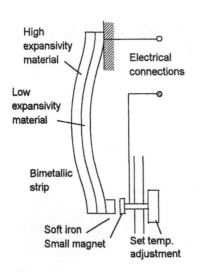

High expansivity material

Low expansivity material

Electrical connections

Bimetallic strip

Soft iron
Small magnet

Set temp. adjustment

Figure 7.1 *Bimetallic thermostat*

7.2 On–off control

With on–off control, the controller is essentially a switch which is activated by the error signal and supplies just an on–off correcting signal. An example is the bimetallic thermostat (Figure 7.1) used with a simple temperature control system. If the actual temperature is above the required temperature, the bimetallic strip is in an off position and the heater is switched off; if the actual temperature is below the required

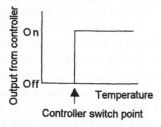

Figure 7.2   *On–off control*

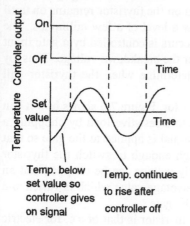

Figure 7.3   *Fluctuation of temperature about set value*

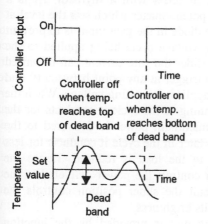

Figure 7.4   *On–off controller with a dead band*

temperature, the bimetallic strip moves into the on position and the heater is switched on. The controller output can thus be just on or off and the correcting signal on or off (Figure 7.2).

Because the control action is discontinuous and there are time lags in the system, oscillations of the controlled variable occur about the required condition. Thus, with temperature control using the bimetallic thermostat, when the room temperature drops below the required level there is a significant time before the heater begins to have an effect on the room temperature and, in the meantime, the temperature has fallen even more. When the temperature rises to the required temperature, since time elapses before the control system reacts and switches the heater off and it cools and stops heating the room, the room temperature goes beyond the required value. The result is that the room temperature oscillates above and below the required temperature (Figure 7.3).

There is also a problem with the simple on–off system in that when the room temperature is hovering about the set value the thermostat might be reacting to very slight changes in temperature and almost continually switching on or off. Thus, when it is at its set value a slight draught might cause it to operate. This problem can be reduced if the heater is switched on at a lower temperature than the one at which it is switched off (Figure 7.4). The term *dead band* is used for the values between the on and off values. For example, if the set value on a thermostat is 20°C, then a deadband might mean it switches on when the temperature falls to 19.5° and off when it is 20.5°. The temperature has thus to change by one degree for the controller to switch the heater on or off and thus smaller changes do not cause the thermostat to switch. A large dead band results in large fluctuations of the temperature about the set temperature; a small dead band will result in an increased frequency of switching. The bimetallic thermostat shown in Figure 7.1 has a permanent magnet on one switch contact and a small piece of soft iron on the other; this has the effect of producing a small dead band in that, when the switch is closed, a significant rise in temperature is needed for the bimetallic element to produce sufficient force to separate the contacts.

On–off control is not too bad at maintaining a constant value of the variable when the capacitance of the system is very large, e.g. a central heating system heating a large air volume, and so the effect of changes in, say, a heater output results in slow changes in the variable. It also involves simple devices and so is fairly cheap. On–off control can be implemented by mechanical switches such as bimetallic strips or relays with more rapid switching being achieved with electronic circuits, e.g. thyristors or transistors used to control the speed of a motor.

7.2.1 Electronic switching

A junction diode has a low resistance to current flow in one direction and a high resistance for the reverse direction. The *thyristor* or *silicon controlled rectifier* (SCR) can be considered to be a diode which can be switched on to be conducting, i.e. switched from having a low resistance to a high resistance, at a particular forward direction voltage. The

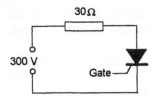

Figure 7.5   *Thyristor circuit*

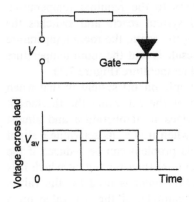

Figure 7.6   *Thyristor d.c. control*

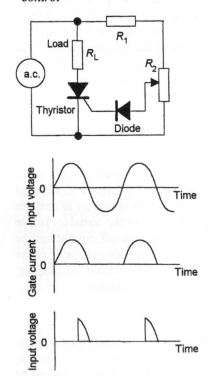

Figure 7.7   *Thristor control for a.c. power to a load*

thyristor passes negligible current when reverse biased and when forward biased the current is also negligible until the forward breakdown voltage, e.g. 300 V, is exceeded. Thus, if such a thyristor is used in a circuit in series with a resistance of 30 Ω (Figure 7.5), before breakdown we have a very high resistance in series with the 30 Ω and so virtually all the 300 V is across the thyristor with its high resistance and there is negligible current. When forward breakdown occurs, the resistance of the thyristor drops to a low value and now, of the 300 V, only about 2 V might be dropped across the thyristor. There is now 300 − 2 = 298 V across the 30 Ω resistor and so the current rises from its negligible value to 298/30 = 9.9 A. When once switched on the thyristor remains on until the forward current is reduced to below a level of a few milliamps. The voltage at which forward breakdown occurs is controlled by a gate input current, the higher the current the lower the breakdown voltage. Thus, by controlling the gate current we can determine when the thyristor will switch from a high to low resistance.

As an illustration of the use of a thyristor, Figure 7.6 shows how it can be used to control the power supplied to a resistive load by chopping a d.c. voltage V. An alternating current signal is applied to the gate so that periodically the voltage V becomes high enough to switch the thyristor off and so the voltage V off. The supply voltage can be chopped and an intermittent voltage produced with an average value which is varied and controlled by the alternating signal to the gate.

Another example of control using a thyristor is that of a.c. for electric heaters, electric motors or lamp dimmers. Figure 7.7 shows a circuit that can be used. The alternating current is applied across the load, e.g. the lamp for a lamp dimming circuit, in series with a thyristor. R_1 is a current-limiting resistor and R_2 is a potentiometer which sets the level at which the thyristor is triggered. The diode in the gate input is to prevent the negative part of the alternating voltage cycle being applied to the gate. By moving the potentiometer slider the gate current can be varied and so the thyristor can be made to trigger at any point between 0° and 90° in the positive half-cycle of the applied alternating voltage. When the thyristor is triggered near the beginning of the cycle it conducts for the entire positive half-cycle and the maximum power is delivered to the load. When triggering is delayed to later in the cycle it conducts for less time and so the power delivered to the load is reduced. Hence the position of the potentiometer slider controls the power delivered to the load; with the light dimming circuit the slider position controls the power delivered to the lamp and so its brightness.

Another form of electronic switching is provided by the junction transistor. For the junction transistor in the circuit shown in Figure 7.8(a), when the base current I_B is zero both the base-emitter and the base-collector junctions are reverse biased. When the base current I_B is increased to a high enough value the base-collector junction becomes forward biased. By switching the base current between 0 and such a value, bipolar transistors can be used as switches. When there is no input voltage V_{in} then virtually the entire V_{CC} voltage appears at the output as the resistance between the collector and emitter is high. When the input voltage is made sufficiently high so that the resistance between the

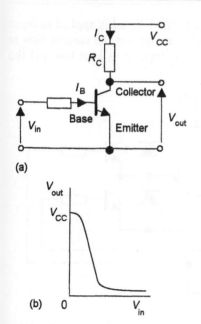

(a)

(b)

Figure 7.8 *Transistor switch*

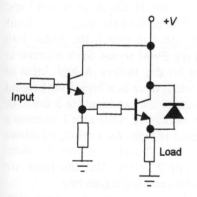

Figure 7.9 *Switching using a Darlington pair*

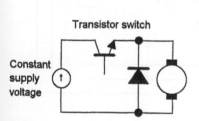

Figure 7.11 *PWM control circuit*

collector and emitter drops to a low value, the transistor switches so that very little of the V_{CC} voltage appears at the output (Figure 7.8(b)). We thus have an electronic switch.

Because the base current needed to drive a bipolar power transistor is fairly large, a second transistor is often needed to take the small current and produce a large enough current to the base of the transistor used for the switching and so enable switching to be obtained with the relatively small currents supplied, for example, by a microprocessor. Such a pair of transistors (Figure 7.9) is termed a *Darlington pair* and they are available as single-chip devices. Since such a circuit is often used with inductive loads and large transient voltages can occur when switching occurs, a protection diode is generally connected in parallel with the switching transistor to prevent damage to it when it is switched off. As an indication of what is available, the integrated circuit ULN2001N contains seven separate Darlington pairs, each pair being provided with a protection diode.

Open-loop control of d.c. motor speed can be achieved by *pulse-width modulation* (PWM). This technique involves the switching on and off of a d.c. voltage to control its average value (Figure 7.10). The greater the fraction of a cycle that the d.c. voltage is switched on the closer its average value is to the input voltage. Figure 7.11 shows how pulse width modulation can be achieved by means of a basic transistor circuit. The transistor is switched on and off by means of a signal applied to its base, e.g. the signal from a microprocessor as a sequence of pulses. By varying the time for which the transistor is switched on so the average voltage applied to the motor can be varied and its speed controlled. Because the motor when rotating acts as a generator, the diode is used to provide a path for the current which arises when the transistor is off.

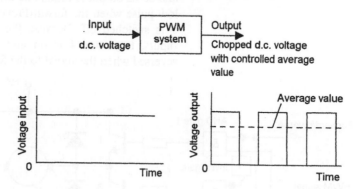

Figure 7.10 *Pulse-width modulation*

Such a basic circuit can only drive the motor in one direction. A circuit (Figure 7.12) involving four transistors, in what is termed an H-circuit, can be used to control both the direction of rotation of the motor and its speed. The motor direction is controlled by which input receives the PWM voltage. In the forward speed motor mode, transistors 1 and 4 are on and current flow is then from left-to-right through the motor. Thus input B is kept low and the PWM signal is applied to input A. For reverse speed, transistors 2 and 3 are on and the current flow is from right-to-left through the motor. Thus input A is kept low and the PWM signal is applied to input B.

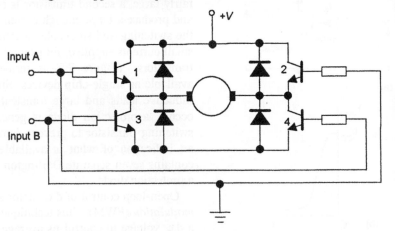

Figure 7.12   *H circuit*

Figure 7.13 shows a better version of the H circuit in which logic gates are used to control inputs A and B to achieve the above conditions with now one input supplied with a signal to switch the motor into forward or reverse and the other input the PWM signal. Such a circuit is better suited to microprocessor control for d.c. motors. A high input to the forward/reverse input means that when there is a high PWM signal the AND gate 1 puts transistor 1 on because the two inputs to it are high and so its output is high. The inverter means that AND gate 2 receives a low pulse when the forward/reverse input is high. As a result, transistor 3 is switched off. Because the AND gates 3 and 4 receive the same inputs, transistor 4 is on and transistor 2 is off. The situations are reversed when the signal to the forward/reverse input goes low.

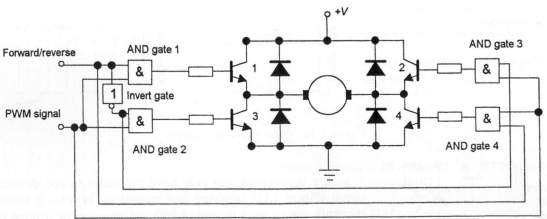

Figure 7.13   *Circuit for microprocessor control of a motor*

The above methods of speed control using PWM have been open-loop systems with the speed being determined by the input to the system and no feedback to modify the input in view of changing load conditions. For a higher grade of speed control than is achieved by the open-loop system, feedback is required. This might be provided by coupling a tacho-generator to the drive shaft; a tachogenerator gives a voltage which is proportional to the rotational speed of the motor. This voltage can be compared with the input voltage used to set the required speed and, after amplification, the error signal used to control the speed of the motor. Figure 7.14 shows how such a closed-loop system might appear when a microprocessor is used as the controller. The analogue output from the tachogenerator is converted to a digital signal by an analogue-to-digital converter. The microprocessor is programmed to compare the digital feedback signal with the set value and give an output based on the error. This output can then be used to control a PWM circuit and so supply a d.c. signal to the motor to control its speed.

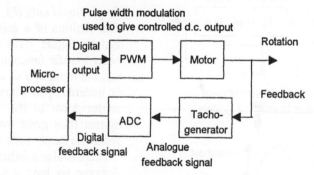

Figure 7.14   *Microprocessor controller with feedback*

7.3 PID control

There are three basic modes of control:

1   *Proportional (P)*
    The controller produces a control action that is proportional to the error e (Figure 7.15), i.e. controller output $= K_p e$ with K_p being the proportional gain.

2   *Derivative (D)*
    The controller produces a control action that is proportional to the rate at which the error is changing de/dt (Figure 7.16), i.e. controller output $= K_d (de/dt)$ with K_d being the derivative gain.

3   *Integral (I)*
    The controller produces a control action that is proportional to the integral of the error e with time (Figure 7.17), i.e. controller output $= K_I \int e \, dt$ with K_I being the integral gain. The integral of the error with time is the total area under the error–time graph up to the time concerned and thus can have a value even when the error has changed back to zero.

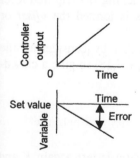

Figure 7.15   *Proportional mode: controller output proportional to error*

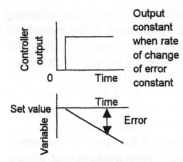

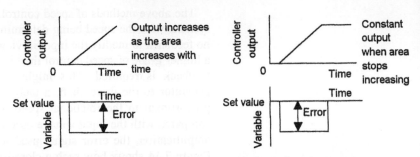

Figure 7.16 *Derivative mode: controller output proportional to rate of change of error, i.e. slope of error–time graph*

Figure 7.17 *Integral mode:controller output proportional to area under error–time graph and does not drop to zero when the error does*

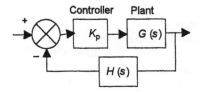

Figure 7.18 *Proportional controller in control system*

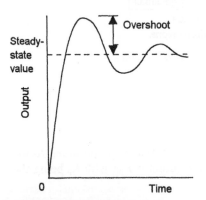

Figure 7.19 *Step response of an under-damped system*

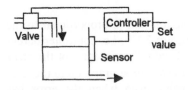

Figure 7.20 *Water level control system*

Controllers tend to be operated using the above modes in the following ways:

1   *Proportional only (P)*
The addition of a gain element to a system can give a closed-loop system (Figure 7.18) which can often be represented by a second-order transfer function. When we have a step input to such a system, e.g. the input of a signal to set the value required, then if the system is underdamped there will be oscillations of the output before it settles down to the steady-state output value (Figure 7.19). The system thus gives overshoot and also takes a significant time to reach the steady state value.

There is also another problem with just using proportional control. Suppose we have a water level control for a tank (Figure 7.20) in which we control the water entering the tank by means of a valve. If there is a change in the outflow from the tank then to maintain the water level at its set value requires the controller to change the signal to the valve to a new value. Because a proportional controller gives an output proportional to the error, there will be no controller output when there is no error. Thus, to maintain the new inflow rate requires the controller to have an error input. So the system will operate with the water level never quite reaching the required level in order to maintain the input flow. This is termed the *offset* or *steady-state error* (Figure 7.21).

We can see the factors affecting the offset by considering the addition of gain K_p to a system (as in Figure 7.18) to give a closed-loop transfer function of:

$$\frac{G(s)K_p}{1 + G(s)H(s)K_p}$$

Suppose we have a simple system where $G(s)$ is just a gain K and $H(s)$ is 1. The output of the system will then be:

$$\text{output} = \frac{KK_p}{1 + KK_p} \times \text{input}$$

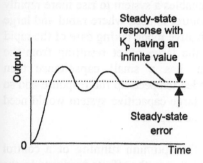

Figure 7.21 *Steady-state error with a step-input to a system*

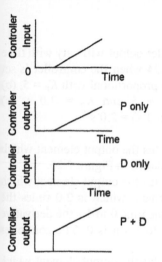

Figure 7.22 *PD controller output to a ramp input*

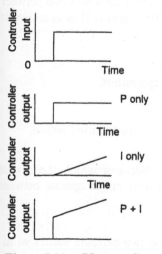

Figure 7.23 *PI controller output to a step input*

As long as KK_p is much greater than 1, the output will become virtually the same as the input. The *offset* or *steady-state error* is the difference between the input and the output in the steady state, the presence of an offset meaning that the system can never give the required output as set by the input (Figure 7.21). Increasing K to minimise the offset will, however, result in an increase in the frequency of the oscillatory output signal and a reduction in the damping, hence an increase in the overshoot. There is thus a problem with just using the proportional mode of control in that to minimise the offset means increasing the overshoot and so a compromise has to be reached.

2 *Proportional plus derivative mode (PD)*

The addition of a derivative mode to a proportional controller modifies its response to inputs. Figure 7.22 shows the response to a ramp input. A PD controller provides an element to the response which is largest when the rate of change of the error is greatest and diminishes as it becomes smaller. Thus with a step input, the controller output rises faster when we apply the step input signal than with just proportional control. With PD control, the output rises more rapidly towards the steady-state value and the overshoot is reduced. Because the derivative mode reduces system oscillations, we can increase the proportional gain element to higher values than would be feasible with just the proportional mode without the oscillations becoming too great a problem and so reduce the offset. The derivative mode is never used alone because it is not capable of maintaining a control signal under steady error conditions. It is always used with the proportional mode and often additionally with the integral mode.

3 *Proportional plus integral (PI)*

The addition of an integral element to a proportional controller modifies the response (Figure 7.23) removing the offset and giving a steady-state value the same as the input set value. This elimination of offset is because the integral mode gives a controlling response which is proportional to the area under the error–time graph up to the current point and so can give a controller output signal even when the error has become zero.. There are many situations where we require a controller to continue giving an output signal even when the error is zero. For example, with the water level control system of Figure 7.20, if the outflow changes to a new rate then the controller has to receive an error signal to maintain the water level constant and still give the new flow rate. With a PI controller, we can still have a controller output with zero error and so there is no need for offset. PI control tends to be used with systems where load disturbances occur frequently.

4 *Proportional plus integral plus derivative (PID)*

The term three-term *controller* is used for PID control. The addition to the proportional mode of the integral mode removes the offset and gives a steady-state value the same as the input set value; the

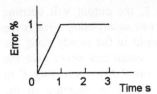

Figure 7.24   *Example*

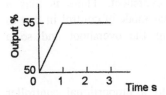

Figure 7.25   *Example*

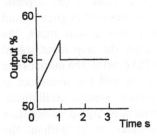

Figure 7.26   *Example*

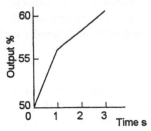

Figure 7.27   *Example*

addition of the derivative mode enables a system to rise more rapidly to the steady-state value. PID control is used where rapid and large disturbances may occur, the derivative mode taking care of the rapid change and the integral mode the large offset resulting from the large disturbance. If a system has a small capacitance then derivative action may not be needed to speed up the response and so PI control might be adequate, a large capacitive system would need the derivative mode to speed up the response.

What has to be determined, for the optimum running of a control system, is the balance to be achieved between the effects produced by the three mode elements. This is called *tuning* the system and is considered later in this chapter in Section 7.7.

Example

Sketch graphs showing how the controller output will vary with time for the error signal shown in Figure 7.24 when the controller is set initially at 50% and operates as (a) just proportional with $K_p = 5$, (b) proportional plus derivative with $K_p = 5$ and $K_d = 2.0$ s, (c) proportional plus integral with $K_p = 5$ and $K_I = 2.0$ s^{-1}.

(a) The controller output will be 50% plus the output element which is 5 times the error signal and so is as shown in Figure 7.25.
(b) The controller output will be 50% plus the output element which is 5 times the error signal and the element which is 2.0 times the slope of the error–time graph. During the time 0 to 1 s the derivative element is 2.0 × 1. During the time 1 to 3 s it is 0. The controller output is thus as shown in Figure 7.26.
(c) The controller output will be 50% plus the output element which is 5 times the error signal and the element which is 2.0 times the area under the error–time graph. At 1.0 s the area has increased from 0 to 0.5. Between 1 and 2 s the area increases by 1 and between 2 and 3 s it increases by 1. The controller output is thus as shown in Figure 7.27.

7.4 Terminology   The following are terms used in describing controllers:

1   *Set-point*
    This is the input of the signal representing the required output.

2   *Range*
    The range is the two extreme values between which the system operates, e.g. a temperature control system might operate between 0°C and 30°C.

3   *Span*
    The span is the difference between the two extreme values within which the system operates, e.g. a temperature control system might operate between 0°C and 30°C and so have a span of 30°C.

4   *Deviation*

The set-point is compared to the measured value to give the deviation or error signal. The term *absolute deviation* is used when the deviation is just quoted as the difference between the measured value and the set value, e.g. a temperature control system might operate between 0°C and 30°C with an absolute deviation of 3°C. This deviation is often quoted as a *fractional* or *percentage deviation*, this being the absolute deviation as a fraction or percentage of the span. Thus, a temperature control system operating between 0°C and 30°C with an error of 3°C has a percentage deviation of $(3/30) \times 100 = 10\%$. When there is no deviation then the percentage deviation is 0% and when the deviation is the maximum permitted by the span it is 100%.

In discussing process control systems it is customary to talk in terms of percentages. Thus percentage deviations are used and items such as valves are discussed in terms of being 50% open.

Example

For the water level control system described in Figure 7.20, the water level is at the required height when the linear control valve has a flow rate of 5 m³/h and the outflow is 5 m³/h. The controller output is then 50% and operates as a proportional controller with a gain of 10. What will be the controller output and the offset when the outflow changes to 6 m³/h?

Since a controller output of 50% corresponds to 5 m³/h from the linear control valve, then 6 m³/h means that the controller output will need to be 60%. To give a change in output of 60 − 50 = 10% with a controller having a gain of 10 means that the error signal into the controller must be 1%. There is thus an offset of 1%.

7.4.1 Proportional band

Generally with process controllers, the proportional gain is described in terms of its *proportional band* (PB). The proportional band is the fractional or percentage deviation that will produce a 100% change in controller output (Figure 7.28):

$$\%PB = \frac{\% \text{ deviation}}{\% \text{ change in controller output}} \times 100$$

A common controller output range is 4 to 20 mA. The 100% controller output might be a signal that fully opens a valve, the 0% being when it fully closes it. A 50% proportional band means that a 50% deviation will produce a 100% change in controller output; 100% proportional band means that a 100% deviation will produce a 100% change in controller output.

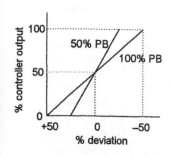

Figure 7.28   *Proportional band*

Since the percentage deviation is the error e as a percentage of the span and the percentage change in the controller output is the controller output y_c as a percentage of the output span of the controller:

$$\%PB = \frac{e}{\text{measurement span}} \times \frac{\text{controller output span}}{y_c} \times 100$$

Since the controller gain K_p is y_c/e:

$$\%PB = \frac{1}{K_p} \frac{\text{controller output span}}{\text{measurement span}} \times 100$$

Example

What is the controller gain of a temperature controller with a 60% PB if its input range is 0°C to 50° and its output is 4 mA to 20 mA?

$$\%PB = \frac{1}{K_p} \frac{\text{controller output span}}{\text{measurement span}} \times 100$$

and so:

$$K_p = \frac{1}{60} \frac{20-4}{50-0} \times 100 = 0.53 \text{ mA/°C}$$

7.5 A process controller

Figure 7.29 shows the basic elements that tend to figure on the front face of a typical three-term process controller. The controller can be operated in three modes by pressing the relevant key:

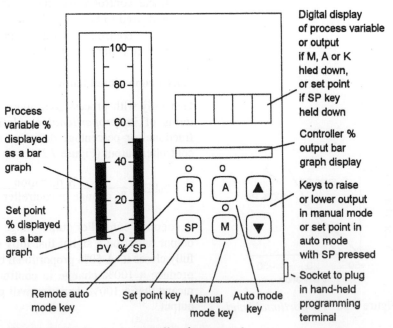

Figure 7.29 *Typical controller front panel*

1   *Manual mode*

The operator directly controls the operation and can increase or decrease the controller output signal by holding down the M key and pressing the up or down keys. A LED above the key shows when this mode has been selected. The output is shown on the digital display and on the bar graph display.

2   *Automatic mode*

The controller operates as a three-term controller with a set point specified by the operator. A LED above the key shows when this mode has been selected. The digital display shows the set point value when the SP key is depressed and the value changed by pressing the up or down keys. The digital display shows the set point value in units such as °C, the unit previously having been set up to give such values in the set up procedure. The set point is also displayed on the vertical bar graph as a percentage.

3   *Remote automatic mode*

The controller is operated in a similar manner to the automatic mode but with the set point established by an external signal. A LED above the key shows when this mode has been selected.

When no key is depressed, the process variable is shown on the digital display and on the vertical bar graph.

The procedure adopted when using the controller is to initially set the mode as manual. The set point is then set to the required value and the controller output manually adjusted until the deviation is zero and the plant thus operating at the required set point. Figure 7.30 shows the block diagram of the control system when it is being operated in manual mode and the operator adjusting the controller output by adding in a signal. The controller can then be switched to automatic control. When this happens, the manual input signal is held constant at the value that was set in manual mode.

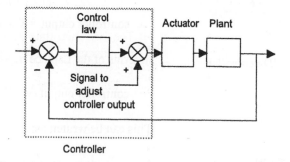

Figure 7.30   *Control system in manual mode*

Switching back to manual mode from automatic mode to make some adjustment and then back to automatic mode gain can present a problem. There can be a sudden change in controller output on the transition from manual to automatic modes, this being termed a 'bump' in the plant operation. This arises because of the integral element in the controller which bases its error on the duration of the error signal input to the controller and does not take account of any manually introduced signals. Thus, changing the manually introduced signal can lead to the output from the controller in the automatic mode not being the same as that in the manual mode. To avoid this 'bump' and give a *bumpless transfer*, modern controllers automatically adjust the contribution to the control law from the integral element.

Modern process controllers are likely to be microprocessor-based controllers, though operating as though they are conventional analogue controllers. They can be programmed by connecting a hand-held terminal to it so that the parameters of the PID controller can be set.

7.6 Controller mathematics

With proportional control (Figure 7.31), the controller produces a control action that is proportional to the error. There is a constant gain K_p acting on the error signal e and so:

$$\text{controller output} = K_p e$$

With derivative control, the controller produces a control action that is proportional to the rate at which the error is changing. Derivative control is not used alone but always in conjunction with proportional control and, often, integral control. Figure 7.32 shows the basic form of a proportional plus derivative controller. The proportional element has an input of the error e and an output of $K_p e$. The derivative element has an input of e and an output which is proportional to the derivative of the error with time, i.e.

$$K_d \frac{de}{dt}$$

where K_d is the derivative gain. Thus the controller output is:

$$\text{controller output} = K_p e + K_d \frac{de}{dt}$$

In terms of the Laplace transform we have:

$$\text{controller output } (s) = \left(K_p + K_d s \right) E(s)$$

This can be written as:

$$\text{controller output } (s) = K_d \left(s + \frac{1}{T_d} \right) E(s)$$

where $T_d = K_d / K_p$ and is called the *derivative time constant*.

With PI control, the proportional element is augmented with an additional element (Figure 7.33) which gives an output proportional to the integral of the error with time. The proportional element has an input of the error e and an output of $K_p e$. The integral element has an input of

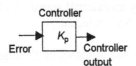

Controller

Figure 7.31 *Proportional controller*

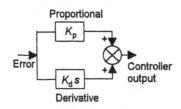

Figure 7.32 *PD controller*

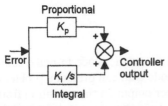

Figure 7.33 *PI controller*

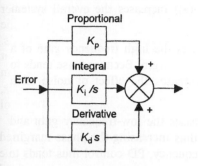

Figure 7.34 *PID controller*

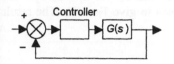

Figure 7.35 *Example*

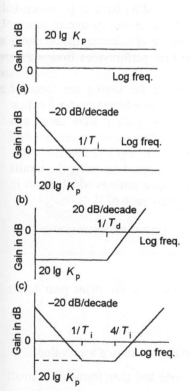

Figure 7.36 *Bode gain plot for (a) P, (b) PI, (c) PD, (d) PID*

e and an output which is proportional to the integral of the error with time, i.e.

$$K_i \int e \, dt$$

where K_i is the integrating gain. Thus the controller output is:

$$\text{controller output} = K_p e + K_i \int e \, dt$$

In terms of the Laplace transform:

$$\text{controller output } (s) = \left(K_p + \frac{K_i}{s} \right) E(s)$$

This can be written as:

$$\text{controller output } (s) = \frac{K_p}{s} \left(s + \frac{1}{T_i} \right) E(s)$$

where $T_i = K_p/K_i$ and is called the *integral time constant*.

Figure 7.34 shows the basic form of a PID, i.e. *three-term controller*. The controller output is:

$$\text{output} = K_p e + K_i \int e \, dt + K_d \frac{de}{dt}$$

Taking Laplace transforms gives:

$$\text{output } (s) = K_p \left(1 + \frac{K_i}{K_p s} + \frac{K_d}{K_p} s \right) E(s)$$

$$= K_p \left(1 + \frac{1}{T_i s} + T_d s \right) E(s)$$

Example

Determine the open-loop transfer function of the system shown in Figure 7.35 if the controller to be used is PD and has a transfer function of $K_p + K_d s$ and $G(s) = \omega_n^2/(s^2 + 2\zeta\omega_n s)$.

The forward path gain, i.e. the open loop gain, is:

$$(K_p + K_d s)\left(\frac{\omega_n^2}{s^2 + 2\zeta\omega_n} \right)$$

7.6.1 Bode plots

Figure 7.36 shows the basic forms of the Bode plots for P, PI and PD controllers.

1 *Proportional control* (Figure 7.36(a)) increases the overall system gain.

2 *PI control* (Figure 7.36(b)) decreases the high frequency gain of a system and thus decreases the phase margin. Because noise tends to be high frequency, PI control decreases the effect of noise on a system.

3 *PD control* (Figure 7.36(c)) decreases the low-frequency gain and reduces the high frequency gain, thus increasing the phase margin. Because noise tends to be high frequency, PD control thus tends to increase the effect of noise on a system.

There is no standard form of Bode plot for a PID controller since the shape depends on the relative values of T_i and T_d. As indicated in the next section, generally the values are chosen to give $T_i = 4T_d$. The result is then as shown in Figure 7.36(d).

7.7 Tuning

With a control system employing just a proportional controller, the value of K_p has to be selected in order to determine the response of the control system to inputs. With a PI controller, K_p and K_i have to be selected. With a PID controller, the values of K_p, K_i and K_d have to be selected. Such selections determine the response of the control system to inputs. The term *tuning* is used to describe the process of selecting the optimum controller settings in order to obtain the best performance from a PID controller.

The most widely used empirical methods for tuning are those of Ziegler and Nichols. They assumed that the open-loop transfer function can be approximated by a first-order system with a time delay and developed two tuning procedures, one called the *ultimate cycle method* which is based on using results from a closed-loop test and the other, called the *process reaction method*, which is based on using the results from open-loop tests. Both are designed to give settings which result in under-damped transient responses with a decay ratio of ¼.

7.7.1 Ultimate cycle method

The procedure used is:

1 Set the controller to manual operation and the plant near to its normal operating conditions.

2 Turn off all control modes but proportional.

3 Set K_p to a low value, i.e. the proportional band to a wide value.

4 Switch the controller to automatic mode and then introduce a small set-point change, e.g. 5 to 10%.

5 Observe the response.

6 Set K_p to a slightly higher value, i.e. make the proportional band narrower.

7 Introduce a small set-point change, e.g. 5 to 10%.

8 Observe the response.

9 Keep on repeating 6, 7 and 8, until the response shows sustained oscillations which neither grow nor decay. Note the value of K_p giving this condition (K_{pu}) and the period (T_u) of the oscillation.

10 Using Table 7.1, determine the optimum controller settings.

Table 7.1 *Settings from the ultimate cycle method*

Type of controller	K_p	T_i	T_d
P	$0.5K_{pu}$		
PI	$0.45K_{pu}$	$T_u/1.2$	
PID	$0.6K_{pu}$	$T_u/2$	$T_u/8$

7.7.2 Process reaction method

The test procedure is:

1 Open the control loop, generally between the controller and the correction elements, so that no control action occurs.

2 Set the controller to manual mode and the plant to near its normal operating conditions.

3 Apply a small step change to the correction element and record the system response.

The graph of the system response plotted against time is called the *process reaction curve* (Figure 7.37). It shows how the system behaves to a step change in controller output. The test signal is expressed as a percentage change in the correction element and the output as a percentage of the full-scale range. A tangent is drawn to give the maximum gradient of the process reaction curve, the maximum gradient being measured as $R = M/T$. The time between when the test signal starts and this tangent intersects the time axis is termed the lag L. Table 7.2 shows the criteria given by Ziegler and Nichols to determine the controller settings.

Table 7.2 *Settings from the process reaction curve method*

Type of controller	K_p	T_i	T_d
P	P/RL		
PI	$0.9P/RL$	$3.3L$	
PID	$1.2P/RL$	$2L$	$0.5L$

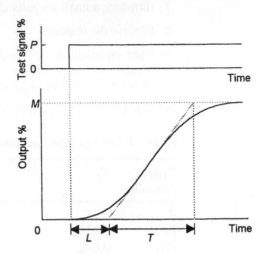

Figure 7.37   *Process reaction curve*

7.8 Digital systems

The term *direct digital control* is used to describe the use of digital computers in the control system to calculate the control signal that is applied to the actuators to control the plant. Such a system is of the form shown in Figure 7.38. At each sample instant the computer samples, via the analogue-to-digital converter (ADC), the plant output to produce the sampled output value. This, together with the discrete input value is then processed by the computer according to the required control law to give the required correction signal which is then sent via the digital-to-analogue converter (DAC) to provide the correcting action to the plant to give the required control. Direct digital control laws are computer programs that take the set value and feedback signals and operate on them to give the output signal to the actuator. The program might thus be designed to implement PID control.

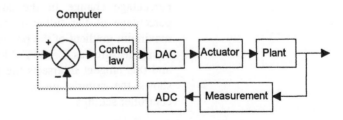

Figure 7.38 *Direct digital control*

The program involves the computer carrying out operations on the feedback measurement value occurring at the instant it is sampled and also using the values previously obtained. The program for proportional control thus takes the form of setting initial values to be used in the program and then a sequence of program instructions which are repeated every sampling period:

Initialise

Set the initial value of the error (this will be zero if the program is to start at the measurement value then occurring)

Set the value of the proportional gain

Loop

Input the error at the instant concerned

Calculate the output by multiplying the error by the set value of the proportional gain

Output the value of the calculated output

Wait for the end of the sampling period

Go back to Loop and repeat the program

For PD control the program is:

Initialise

Set the initial value of the error (this will be zero if the program is to start at the measurement value then occurring)

Set the initial value of the error that is assumed to have occurred in the previous sampling period

Set the value of the proportional gain

Set the value of the derivative gain

Loop

Input the error at the instant concerned

Calculate the proportional part of the output by multiplying the error by the set value of the proportional gain

Calculate the derivative part of the output by subtracting the value of the error at the previous sampling instant from the value at the current sampling instant (the difference is a measure of the rate of change of the error since the signals are sampled at regular intervals of time) and multiply it by the set value of the derivative gain.

Calculate the output by adding the proportional and derivative output elements

Output the value of the calculated output

Wait for the end of the sampling period

Go back to Loop and repeat the program

For PI control the program is:

Initialise

Set the initial value of the error (this will be zero if the program is to start at the measurement value then occurring)

Set the value of the output that is assumed to have occurred in the previous sampling period

Set the value of the proportional gain

Set the value of the integral gain

Loop

Input the error at the instant concerned

Calculate the proportional part of the output by multiplying the error by the set value of the proportional gain

Calculate the integral part of the output by multiplying the value of the error at the current sampling instant by the sampling period and the set value of the integral gain (this assumes that the output has remained constant over the previous sampling period and so multiplying its value by the sampling period gives the area under the output–time graph) and add to it the previous value of the output.

Calculate the output by adding the proportional and integral output elements

Output the value of the calculated output

Wait for the end of the sampling period

Go back to Loop and repeat the program

7.8.1 Programmable logic controllers

A *programmable logic controller* (PLC) is a special form of micro-processor-based controller that uses a programmable memory to store instructions and is designed to be operated by engineers with perhaps a limited knowledge of computers and computing languages. Thus, the designers of the PLC have pre-programmed it so that the control program can be entered using a simple pictorial form of language called *ladder programs*. The term *logic* is used because programming is mainly concerned with implementing logic and switching operations, e.g. if A or B occurs switch on C, if A and B occurs switch on D. For example, it might be used to control the level of water in a tank by a sensor giving an input signal when the tank is empty and another sensor giving a signal when the tank is full. Thus when the tank-empty sensor gives an on input the controller gives an on output signal to open a valve to allow water into the tank. This output remains on until the tank-full sensor gives an input signal, the controller then switches off the output signal to the valve.

Input devices, e.g. sensors such as switches, and output devices in the system being controlled, e.g. motors, valves, etc., are connected to the PLC. The operator then enters a sequence of instructions, i.e. a program, into the memory of the PLC. The controller then monitors the inputs and outputs according to this program and carries out the control rules for which it has been programmed. Many PLCs also can be programmed to operate as PID controllers.

Typically a PLC system has five basic components. These are the processor unit, memory, the power supply unit, input/output interface section and the programming device. Figure 7.39 shows the basic arrangement.

1   The *processor unit* or *central processing unit (CPU)* is the unit containing the microprocessor and this interprets the input signals and carries out the control actions, according to the program stored in its memory, communicating the decisions as action signals to the outputs.

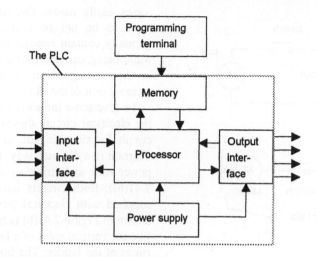

Figure 7.39 *The PLC system*

2 The *power supply unit* is needed to convert the mains a.c. voltage to the low d.c. voltage (5 V) necessary for the processor and the circuits in the input and output interface modules.

3 The *programming device* is used to enter the required program into the memory of the processor. The program is developed in the device and then transferred to the memory unit of the PLC.

4 The *memory unit* is where the program is stored that is to be used for the control actions to be exercised by the microprocessor.

5 The *input and output sections* are where the processor receives information from external devices and communicates information to external devices. The inputs might be from switches, temperature sensors, or flow sensors, etc. combined with appropriate signal processing elements. The outputs might be to motor starter coils, solenoid valves, etc.

Programs are entered into a PLC's memory using a program device which is usually not permanently connected to a particular PLC and can be moved from one controller to the next without disturbing operations. For the operation of the PLC it is not necessary for the programming device to be connected to the PLC since it transfers the program to the PLC memory. Programming devices can be a hand-held device, a desktop console or a computer. Hand-held systems incorporate a small keyboard and liquid crystal display, Figure 7.40 showing a typical form. Desktop devices are likely to have a visual display unit with a full keyboard and screen display. Personal computers are widely configured as program development workstations. Some PLCs only require the computer to have appropriate software, others special communication cards to interface with the PLC. A major advantage of using a computer is that the program can be stored on the hard disk or a floppy disk and

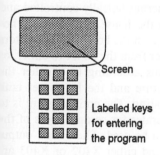

Figure 7.40 *Hand-held programmer*

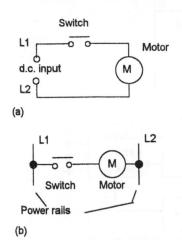

(a)

(b)

Figure 7.41   *A motor switching circuit: (a) conventional form of circuit, (b) ladder form of circuit*

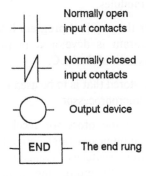

Figure 7.42   *Basic symbols used with ladder programs*

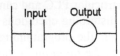

Figure 7.43   *A ladder rung*

copies easily made. The disadvantage is that the programming often tends to be not so user-friendly. Hand-held programming consoles normally contain enough memory to allow the unit to retain programs while being carried from one place to another. Only when the program has been designed on the programming device is it transferred to the memory unit of the PLC.

To give some indication of the ladder form of programming, consider the electrical circuit shown in Figure 7.41(a). The diagram shows the circuit for switching on or off an electric motor. We can redraw this diagram in a different way, using two vertical lines to represent the input power rails and stringing the rest of the circuit between them (Figure 7.41(b)). Both circuits have the switch in series with the motor and supplied with electrical power when the switch is closed. The circuit shown in Figure 7.41(b) is termed a *ladder diagram*. The power lines are like the vertical sides of a ladder with the horizontal circuit lines like the rungs of the ladder. The horizontal rungs show only the control portion of the circuit, in the case of Figure 7.41 it is just the switch in series with the motor. Drawing ladder diagrams is a means of writing programs that is used with PLCs.

Figure 7.42 shows the basic standard ladder program symbols that are used for input and output devices. Note that inputs are represented by just two symbols representing normally open or normally closed contacts. This applies whatever the form of the device connected to the input. The action of the input has to be designed to be equivalent to opening or closing a switch. Outputs are represented by just one symbol, regardless of the device connected to the output. To illustrate the drawing of the rung of a ladder diagram, consider a situation where the energising of an output device, e.g. a motor, depends on a sensor–signal processing arrangement being like a normally open start switch which on being activated is effectively closed, i.e. the input turns from a low signal to a high signal. Figure 7.43 shows the ladder diagram. Starting with the input, we have the normally open symbol ‖ for the input contacts. There are no other input devices and the line terminates with the output, denoted by the symbol O. When the switch is closed, i.e. the input is high, the output of the motor is activated.

As an illustration of the use of a PLC system, consider its use to control the temperature of a domestic central heating system (Figure 7.44). The central heating boiler is to be thermostatically controlled and supply hot water to the radiator system in the house and also to a hot water tank to provide hot water from the taps in the house. Pump motors have to be switched on to direct the hot water from the boiler to either, or both, of the radiator and hot water systems according to whether the temperature sensors for the room temperature and the hot water tank indicate that the radiators or tank need heating. The entire system is to be controlled by a clock so that it only operates for certain hours of the day. Figure 7.45 shows a program that can be used. The boiler, output Y430, is switched on if X400 and X401 and either X402 or X403 are switched on. This means if the clock switched is on, the boiler temperature sensor gives an on input, and either the room temperature sensor or the water temperature sensors give on inputs. The motorised valve M1, output Y431, is switched on if the boiler, Y430, is on and if the room temperature sensor X402 gives an on input. The motorised valve M2, output Y432, is switched on if the boiler, Y430, is on and if the water temperature sensor gives an on input.

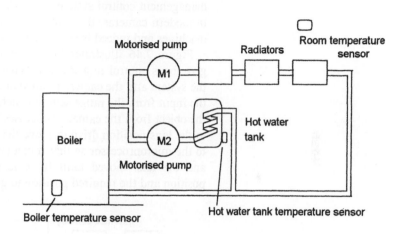

Figure 7.44 *Central heating system*

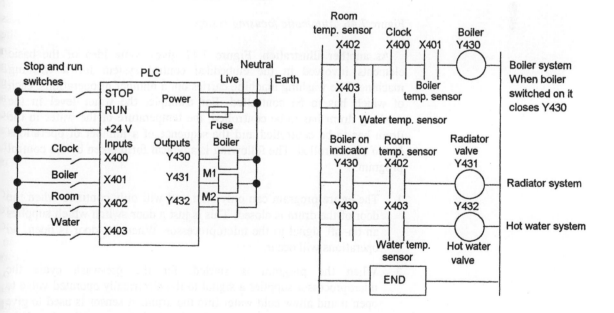

Figure 7.45 *Central heating system with PLC controller*

7.8.2 Embedded systems

The term *embedded system* is used for control systems involving a microprocessor being used as the controller and located as an integral element, i.e. embedded, in the system. Such a system is used with engine management control systems in modern cars, exposure and focus control in modern cameras, the controlling of the operation of modern washing machines and indeed is very widely used in modern consumer goods.

Figure 7.46 illustrates how an embedded microprocessor is used to give focus control in a camera. When the switch is operated to activate the system and the camera pointed at an object, the microprocessor takes the input from the range sensor which gives a measure of the distance of the object from the camera, processes it and gives an output which is fed to the lens position drive to move the lens. The lens position is fed back to the microprocessor so that it can be compared with the required value and the lens moved until there is no error between the actual lens position and the required position to give a focused image.

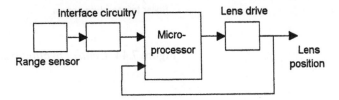

Figure 7.46 *Automatic focusing system*

As another illustration, Figure 7.47 gives some idea of the basic elements involved in the embedded control system for a washing machine. The washing machine carries out a number of operations, each of which has to be controlled. For example, the water level in the washing drum has to be controlled, the temperature of the water in the drum has to be controlled and the sequence of a number of operations has to be controlled. The following is a typical form taken by the control program:

1   The entire program can only start and will only continue when the door of the drum is closed. This is just a door switch which supplies an on–off signal to the microprocessor. When the door is open, no operations will occur.

2   When the program is started, for the pre-wash cycle the microprocessor supplies a signal to the electrically operated valve to open it and allow cold water into the drum. A sensor is used to give a signal when the water level has reached the preset level and the microprocessor gives the output to switch off the current to the valve.

3   After a predetermined time, the microprocessor supplies a signal to a pump to empty the drum.

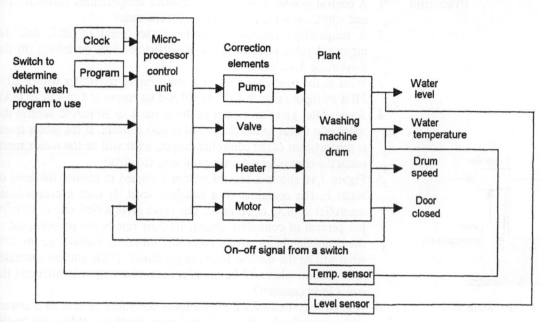

Figure 7.47 *Washing machine system*

4   For the main wash cycle, the microprocessor gives an output signal which again opens the valve to allow cold water into the drum. This level is sensed and the water shut off when the required level is reached.

5   The microprocessor then supplies a signal to switch on the current to an electric heater to heat the water. A temperature sensor sends a signal back to the microprocessor and it switches off the current when the water temperature reaches the preset value.

6   The microprocessor then supplies a signal to switch on the drum motor to rotate the drum. This might be open-loop control and just continue for the time determined by the microprocessor.

7   Then the microprocessor switches off the motor and supplies a signal to a discharge pump to empty the water from the drum.

8   The rinse part of the operation is now switched as a sequence of signals to open valves which allow cold water into the machine to a predetermined level, switch it off, operate the motor to rotate the drum, operate a pump to empty the water from the drum, and repeat this sequence a number of times.

9   The final part of the operation is when the microprocessor supplies a signal to switch on the motor, at a higher speed than for the rinsing, to spin the clothes.

Problems

1 A control system is designed to control temperatures between $-10°$ and $+30°C$. What is (a) the range, (b) the span?

2 A temperature control system has a set point of $20°C$ and the measured value is $18°C$. What is (a) the absolute deviation, (b) the percentage deviation?

3 What is the controller gain of a temperature controller with a 80% PB if its input range is $40°C$ to $90°$ and its output is 4 mA to 20 mA?

4 A controller gives an output in the range 4 to 20 mA to control the speed of a motor in the range 140 to 600 rev/min. If the motor speed is proportional to the controller output, what will be the motor speed when the controller output is (a) 8 mA, (b) 40%?

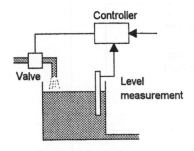

Controller

Valve

Level measurement

Figure 7.48   *Problem 5*

5 Figure 7.48 shows a control system designed to control the level of water in the container to a constant level. It uses a proportional controller with K_p equal to 10. The valve gives a flow rate of 10 m^3/h per percent of controller output, its flow rate being proportional to the controller input. If the controller output is initially set to 50% what will be the outflow from the container? If the outflow increases to 600 m^3/h, what will be the new controller output to maintain the water level constant?

6 A control system uses a proportional controller to control a system with a transfer function of K and unity feedback. What will be the offset error if the proportional controller has a gain K_p of 10 and $K = 0.3$ and a step input of 4 units is applied?

7 A control system uses a proportional controller to control a system with a transfer function of K and unity feedback. What should the gain K_p of the controller be to give an offset error of 0.01 unit if $K = 0.1$ and there is a step input of 4 units to the system?

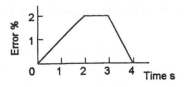

Figure 7.49   *Problem 8*

8 Sketch graphs showing how the controller output will vary with time for the error signal shown in Figure 7.49 when the controller is set initially at 50% and operates as (a) just proportional with $K_p = 5$, (b) proportional plus derivative with $K_p = 5$ and $K_d = 1.0$ s, (c) proportional plus integral with $K_p = 5$ and $K_I = 0.5$ s^{-1}.

9 Using the Ziegler–Nichols ultimate cycle method for the determination of the optimum settings of a PID controller, oscillations began with a 30% proportional band and they had a period of 11 min. What would be the optimum settings for the PID controller?

10 Using the Ziegler–Nichols ultimate cycle method for the determination of the optimum settings of a PID controller, oscillations began with a gain of 2.2 with a period of 12 min. What would be the optimum settings for the PID controller?

11 Figure 7.50 shows the open-loop response of a system to a unit step in controller output. Using the Ziegler–Nichols data, determine the optimum settings of the PID controller.

12 A closed loop control system has a PID controller with transfer function $K_p + (K_i/s) + K_d s$ and is cascaded with a process having a transfer function of $10/(s + 5)(s + 10)$. If the system has unity feedback, what is the transfer function of the closed-loop system?

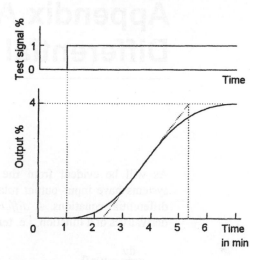

Figure 7.50 *Problem 11*

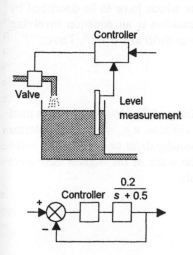

Figure 7.51 *Problem 14*

13 A closed loop control system has unity feedback and a plant with transfer function $100/[s(s + 0.1s)(1 + 0.2s)]$. By drawing the Bode diagrams, determine the phase margin when the following controllers are used: (a) a proportional controller with transfer function 1, (b) a PD controller with transfer function $1 + 0.5s$.

14 Figure 7.51 shows a liquid level control system and its representation by a block diagram. Determine the way the output will vary with time if the controller is (a) proportional only with a proportional gain of 2, (b) integral only with an integral gain of 2.

Appendix A
Differential equations

As will be evident from the models discussed in Chapter 2, many systems have input–output relationships which have to be described by differential equations. A *differential equation* is an equation involving derivatives of a function, i.e. terms such as dy/dt and d^2y/dt^2. Thus:

$$\tau \frac{dy}{dt} + y = 0$$

is a differential equation. The term *ordinary differential equation* is used when there are only derivatives of one variable, e.g. we only have terms such as dy/dt and d^2y/dt^2 and not additionally dx/dt or dx^2/dt^2. The *order* of a differential equation is equal to the order of the highest derivative that appears in the equation. For example,

$$\tau \frac{dy}{dt} + y = 0$$

and

$$\tau \frac{dy}{dt} + y = kx$$

are first-order ordinary differential equations since the highest derivative is dy/dt and there are derivatives of only one variable. The first of the above two equations is said to be *homogeneous* since it only contains terms involving y. Such an equation is given by a system which has no forcing input; one with a forcing input gives a non-homogeneous equation. For example, an electrical circuit containing just a charged capacitor in series with a resistor will give a homogeneous differential equation describing how the potential difference across the capacitor changes with time as the charge leaks off the capacitor; there is no external source of voltage. However, if we have a voltage source which is switched into the circuit with the capacitor and resistor then the voltage source gives a forcing input and a non-homogeneous equation results.

The equations:

$$m \frac{d^2y}{dt^2} + c \frac{dy}{dt} + ky = 0$$

and

$$m\frac{d^2y}{dt^2} + c\frac{dy}{dt} + ky = F$$

are examples of a second-order differential equation since the highest derivative is d^2y/dt^2. The first of the two is homogeneous since there is no forcing input and the second is non-homogeneous with a forcing input F.

Solving a first-order differential equation

With a first-order differential equation, if the variables are separable, i.e. it is of the form:

$$\frac{dy}{dt} = f(t)$$

then we can solve such equations by integrating both sides of the equation with respect to x:

$$\int \frac{dy}{dt}\, dt = \int f(t)\, dt$$

This is equivalent to separating the variables and writing:

$$\int dt = \int f(t)\, dt$$

Consider the response of a first-order system to a step input, e.g. the response of a thermometer when inserted suddenly into a hot liquid. This sudden change is an example of a step input. In Chapter 2 the differential equation for such a change was determined as:

$$RC\frac{dT}{dt} + T = T_L$$

where T is the temperature indicated by the thermometer, T_L the temperature of the hot liquid, R the thermal resistance and C the thermal capacitance. We can solve such an equation by the technique of 'separation of variables'. Separating the variables gives:

$$\frac{1}{T_L - T}\, dT = \frac{1}{RC}\, dt$$

Integrating then gives:

$$-\ln (T_L - T) = (1/RC)t + A$$

where A is a constant. This equation can be written as:

$$T_L - T = e^A\, e^{-t/\tau} = B\, e^{-t/\tau}$$

with B being a constant and $\tau = RC$. τ is termed the *time constant* and can be considered to be the time that makes the exponential term e^{-1}. If we consider the thermometer to have been inserted into the hot liquid at time $t = 0$ and to have been indicating the temperature T_0 at that time, then, since $e^0 = 1$, we have $B = T_L - T$. Hence the equation can be written as:

$$T = (T_0 - T_L)\, e^{-t/\tau} + T_L$$

The exponential term will die away as t increases and so gives the transient part of the response. T_L is the steady-state value that will be attained eventually.

Complementary function and particular integral

Suppose we have the first-order differential equation $dy/dt + y = 0$. Such an equation is homogeneous and, if we apply the technique of separation of the variables, has the solution $y = C\, e^{-t}$. Now suppose we have the non-homogeneous equation $dy/dt + y = 2$. If we now use the separation of variable technique we obtain the solution $y = C\, e^{-t} + 2$. Thus, its solution is the sum of the solution for the homogeneous equation plus another term. The solution of the homogeneous differential equation is called the *complementary function* and, when there is a forcing input with that differential equation, the term added to it for the non-homogeneous solution is called the *particular integral*. We can, easily, obtain a particular integral by assuming it will be of the same form as the forcing input. Thus if this is a constant then we try $y = A$, if of the form $a + bx + cx^2 + \dots$ then $y = A + Bx + Cx^2 + \dots$ is tried, if an exponential then $y = A\, e^{kx}$ is tried, if a sine or cosine then $y = A \sin \omega x + B \cos \omega x$ is tried.

As an illustration, consider the system involving solved in the previous section of a thermometer being inserted into a hot liquid and the relationship being given by:

$$RC\frac{dT}{dt} + T = T_L$$

The homogeneous form of this equation is:

$$RC\frac{dT}{dt} + T = 0$$

We can solve this equation by the technique of the separation of variables to give the complementary function:

$$\frac{1}{T}\, dT = -\frac{1}{RC}\, dt$$

Integrating then gives:

$$-\ln T = -(1/RC)t + A$$

where A is a constant. This equation can be written as:

$$-T = e^A\, e^{-t/\tau} = B\, e^{-t/\tau}$$

$$T = -B\,e^{-t/\tau}$$

Now consider the particular integral for the non-homogeneous equation and, because the forcing input is a constant, we try $T = C$. Since $dT/dt = 0$ then the substituting values in the differential equation gives:

$$0 + C = T_L$$

Thus the particular solution is $T = T_L$. Hence the full solution is the sum of the complementary solution and the particular integral and so:

$$T = -B\,e^{-t/\tau} + T_L$$

If we consider the thermometer to have been inserted into the hot liquid at time $t = 0$ and to have been indicating the temperature T_0 at that time, then, since $e^0 = 1$, we have $B = T_L - T$. Hence, as before, the equation can be written as:

$$T = (T_0 - T_L)\,e^{-t/\tau} + T_L$$

As a further illustration, consider the thermometer, in equilibrium in a liquid at temperature T_0, when the temperature of the liquid is increased at a constant rate a, i.e. a so-called ramp input. The temperature will vary with time so that after a time t it is $at + T_0$. This is then the forcing input to the system and so we have the differential equation:

$$RC\frac{dT}{dt} + T = at + T_L$$

The homogeneous form of this equation is:

$$RC\frac{dT}{dt} + T = 0$$

and, as before, the complementary solution is:

$$T = -B\,e^{-t/\tau}$$

For the particular integral we try a solution of the form $T = D + Et$. Since $dT/dt = E$, substituting in the non-homogeneous differential equation gives:

$$RCE + D + Et = aT + T_0$$

Equating all the coefficients of the t terms gives $E = a$ and equating all the non-t terms gives $RCE + D = T_0$ and so $D = T_0 - RCa$. Thus the particular integral is:

$$T = T_0 - RCa + at$$

and so the full solution, with $\tau = RC$, is:

$$T = -B\ e^{-t/\tau} + T_0 - \tau a + at$$

When $t = 0$ we have $T = T_0$ and so $B = -\tau a$. Hence we have:

$$T = \tau a\ e^{-t/\tau} + T_0 - \tau a + at$$

Solving a second-order differential equation

We can use the technique of finding the complementary function and the particular integral to obtain the solution of a second-order differential equation. Consider a spring–damper–mass system of the form shown in Figure 2.8. The differential equation for the displacement y of the mass when subject to step input at time $t = 0$ of a force F is:

$$m\frac{d^2y}{dt^2} + c\frac{dy}{dt} + ky = F$$

In the absence of damping and the force F we have the homogeneous differential equation :

$$m\frac{d^2y}{dt^2} + ky = 0$$

This describes an oscillation which has an acceleration d^2y/dt^2 which is proportional to $-y$ and is a description of simple harmonic motion; we have a mass on a spring allowed to freely oscillate without any damping. Simple harmonic motion has a displacement $y = A \sin \omega t$. If we substitute this into the differential equation we obtain:

$$-mA\omega^2 \sin \omega t + kA\omega \sin \omega t = 0$$

and so $\omega = \sqrt{(k/m)}$. This is termed the *natural angular frequency* ω_n. If we define a constant called the *damping ratio* of:

$$\zeta = \frac{c}{2\sqrt{mk}}$$

then we can write the differential equation as:

$$\frac{1}{\omega_n^2}\frac{d^2y}{dt^2} + \frac{2\zeta}{\omega_n}\frac{dy}{dt} + y = \frac{F}{k}$$

This differential equation can be solved by the method of determining the complementary function and the particular integral. For the homogeneous form of the differential equation, i.e. the equation with zero input, we have:

$$\frac{1}{\omega_n^2}\frac{d^2y}{dt^2} + \frac{2\zeta}{\omega_n}\frac{dy}{dt} + y = 0$$

We can try a solution of the form $y = A\ e^{st}$. This, when substituted, gives:

$$\frac{1}{\omega_n^2}s^2 + \frac{2\zeta}{\omega_n}s + 1 = 0$$

$$s^2 + 2\omega_n\zeta s + \omega_n^2 = 0$$

This equation has the roots:

$$s = \frac{-2\omega_n\zeta \pm \sqrt{4\omega_n^2\zeta^2 - 4\omega_n^2}}{2} = -\omega_n\zeta \pm \omega_n\sqrt{\zeta^2 - 1}$$

When we have:

1   *Damping ratio between 0 and 1*
    There are two complex roots:

$$s = -\zeta\omega_n \pm j\omega_n\sqrt{1 - \zeta^2}$$

If we let:

$$\omega = \omega_n\sqrt{1 - \zeta^2}$$

we can write:

$$s = -\zeta\omega_n \pm j\omega$$

Thus:

$$y = A\,e^{(-\zeta\omega_n + j\omega)t} + B\,e^{-\zeta(\omega_n - j\omega)t}$$

$$= e^{-\zeta\omega_n t}(A\,e^{j\psi t} + B\,e^{-j\omega t})$$

Using Euler's equation, we can write this as:

$$y = e^{-\zeta\omega_n t}(P\cos\omega t + Q\sin\omega t)$$

This can be written in an alternative form.   If we consider a right-angled triangle with angle ϕ with P and Q being opposite sides of the triangle (Figure ApA.1) then $\sin\phi = P/\sqrt{(P^2 + Q^2)}$ and $\cos\phi = P/\sqrt{(P^2 + Q^2)}$. Hence, using the relationship $\sin(\omega t + \phi) = \sin\omega t\cos\phi + \cos\omega t\sin\phi$, we can write:

$$y = C\,e^{-\zeta\omega_n t}\sin(\omega t + \phi)$$

where C is a constant and ϕ a phase difference. This describes a damped sinusoidal oscillation. Such a motion is said to be *under damped*.

2   *Damping ratio equal to 1*
    This gives two equal roots $s_1 = s_2 = -\omega_n$. and the solution:

Figure ApA.1   *Angle ϕ*

$$y = (At + B)\, e^{-\omega_n t}$$

where A and B are constants. This describes an exponential decay with no oscillations. Such a motion is said to be *critically damped*.

3   *Damping ratio greater than 1*
    This gives two real roots:

$$s_1 = -\omega_n \zeta + \omega_n \sqrt{\zeta^2 - 1}$$

$$s_2 = -\omega_n \zeta - \omega_n \sqrt{\zeta^2 - 1}$$

and hence:

$$y = A\, e^{s_1 t} + B\, e^{s_2 t}$$

where A and B are constants. This describes an exponential decay taking longer to reach the steady-state value than the critically damped case. It is said to be *over damped*.

The above analysis has given the complementary functions for the second-order differential equation. For the particular integral, in this case where we have a step input of size F, we can try the particular integral $x = A$. Substituting this in the differential equation gives $A = F/k$ and thus the particular integral is $y = F/k$. Thus the solutions to the differential equation are:

1   *Damping ratio between 0 and 1, i.e. under damped*

$$y = C\, e^{-\zeta \omega_n t} \sin(\omega t + \phi) + F/k$$

2   *Damping ratio equal to 1, i.e. critically damped*

$$y = (At + B)\, e^{-\omega_n t} + F/k$$

3   *Damping ratio greater than 1, i.e. over damped*

$$y = A\, e^{s_1 t} + B\, e^{s_2 t} + F/k$$

In all cases, as t tends to infinite then y tends to the value F/k. Thus the steady-state value is F/k.

Appendix B
Laplace transform

Control systems tend to have input–output relationships which are described by differential equations. The differential equations describe how the output varies with time for an input. We can determine the output for a system with some particular input by solving the differential equation. However, the Laplace transform is a way of transforming differential equations into more convenient forms for determining outputs.

A quantity which is a function of time can be represented as $f(t)$ and is said to be in the *time domain*, e.g. if we have a voltage v which is a function of time we can write as $v(t)$ to show that it is. In discussing control systems we are only concerned with values of time greater than or equal to 0, i.e. $t \geq 0$, and so to obtain the Laplace transform of $f(t)$ function we multiply it by e^{-st} and then integrate with respect to time from zero to infinity; s is a constant with the unit of 1/time. The result is then said to be in the *s-domain*. The Laplace transform of the function of time $f(t)$, which is written as $\mathcal{L}\{f(t)\}$, is thus:

$$\mathcal{L}\{f(t)\} = \int_0^\infty e^{-st} f(t)\, dt$$

A function of s is written as $F(s)$. It is usual to use a capital letter F for the Laplace transform and a lower-case letter f for the time-varying function $f(t)$. Thus:

$$\mathcal{L}\{f(t)\} = F(s)$$

The procedure for using the Laplace transform is to:

1   Transform functions of time into functions of s.

2   Carry out algebraic manipulations in the s-domain. We can carry out algebraic manipulations on a quantity in the s-domain, i.e. adding, subtracting, dividing and multiplying in the way we do with any algebraic quantities.

3   Transform the functions of s back into functions of time, i.e. the so-termed inverse operation. This involves finding the time domain function that could have given the s-domain expression obtained by the algebraic manipulations.

For the inverse operation, when the function of time is obtained from the Laplace transform, we can write

$$f(t) = \mathcal{L}^{-1}\{F(s)\}$$

Obtaining the Laplace transform

The following examples illustrate how we can obtain, from first principles, the Laplace transform of functions of time. However, in practice, engineers use a table of Laplace transforms to avoid having to carry out such calculations; Table 3.1 shows some of the transformations commonly encountered.

The unit step is the type of function encountered when we suddenly apply an input to a system (Figure Ap.B.1). It describes an abrupt change in some quantity from zero to a steady value, e.g. the change in the voltage applied to a circuit when it is suddenly switched on. Thus, at time $t = 0$ it suddenly switches to the value 1 and has the constant value of 1 for all values of time greater than 0, i.e. $f(t) = 1$ for $t \geq 0$. The Laplace transform of the unit step function is:

$$\mathcal{L}\{f(t)\} = F(s) = \int_0^\infty 1\,e^{-st}\,dt = -\frac{1}{s}[e^{-st}]_0^\infty$$

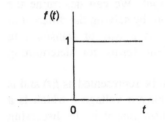

Figure Ap.B.1   *Unit-step function*

For $t = \infty$ the value of the exponential term $e^{-\infty}$ is 0 and with $t = 0$ the value of e^{-0} is -1. Hence:

$$F(s) = \frac{1}{s}$$

As a further illustration, consider the Laplace transform of the function $f(t) = e^{at}$, where a is a constant :

$$F(s) = \int_0^\infty e^{at}e^{-st}\,dt$$

$$= \int_0^\infty e^{-(s-a)t}\,dt = -\frac{1}{s-a}[e^{-(s-a)t}]_0^\infty$$

When $t = \infty$ the term in the brackets becomes 0 and when $t = 0$ it becomes -1. Thus:

$$F(s) = \frac{1}{s-a}$$

Properties of the Laplace transform

To use the Laplace transforms of functions given in tables it is necessary to know the properties of such transforms. The following are basic properties:

1   *Linearity property*
    The Laplace transform of the sum of two time functions, e.g. $f(t)$ and $g(t)$, can be obtained by adding the Laplace transforms of the two separate Laplace transforms, i.e.

$$\mathcal{L}\{f(t) + g(t)\} = \mathcal{L}f(t) + \mathcal{L}g(t)$$

Thus the Laplace transform of $1 + t$ is:

$$\mathcal{L}\{1 + t\} = \mathcal{L}1 + \mathcal{L}t = \frac{1}{s} + \frac{1}{s^2}$$

2   *Multiplication by a constant*

As a consequence of the linearity property, suppose we want the Laplace transform of $2t$. This can be considered to be $t + t$ and so the Laplace transform of $2t$ is the sum of the transforms of t and t and so $2 \times$ the Laplace transform of t. Since the Laplace transform of t is $1/s$ then the Laplace transform of $2t$ is $2/s$. The Laplace transform $4t^2$ is given by 4 times the Laplace transform of t^2 and so, since the Laplace transform of t^2 is $2/s^2$, is $8/s^2$. In general:

$$\mathcal{L}\{af(t)\} = a\mathcal{L}f(t)$$

The Laplace transform of $1 + 2t + 4t^2$ is given by the sum of the transforms of the individual terms in the expression and so is:

$$F(s) = \frac{1}{s} + \frac{2}{s^2} + \frac{8}{s^3}$$

3   *Final-value theorem*

This theorem enables us to determine the value a function of time will end up with after a long period of time, i.e. the steady-state value. The final-value theorem can be stated as: if a function of time $f(t)$ has a Laplace transform $F(s)$ then in the limit as the time tends to infinity the value of the function is given by

$$\lim_{t \to \infty} f(t) = \lim_{s \to 0} sF(s)$$

For example, consider the final value of e^{-3t} as t tends to infinity. Using Table 3.1, we have $e^{-at} = 1/(s + a)$ and so $e^{-3t} = 1/(s + 3)$. The final-value theorem thus gives:

$$\lim_{t \to \infty} f(t) = \lim_{s \to 0} sF(s) = \lim_{s \to \infty} \frac{s}{s+3} = 0$$

and so the steady-state value is 0.

4   *Derivatives*

The Laplace transform of a derivative of a function $f(t)$ is given by:

$$\mathcal{L}\left\{\frac{\mathrm{d}}{\mathrm{d}t}f(t)\right\} = sF(s) - f(0)$$

where $f(0)$ is the value of the function when $t = 0$. For a second derivative:

$$\mathcal{L}\left\{\frac{d^2}{dt^2}f(t)\right\} = s^2F(s) - sf(0) - \frac{d}{dt}f(0)$$

where $df(0)/dt$ is the value of the first derivative at $t = 0$.

For example, if we have a function of time with an initial zero value, i.e. $f(0) = 2$, then the Laplace transform of $df(t)/dt$ is $sF(s)$. If the function had $f(0) = 2$ then the Laplace transform of $df(t)/dt$ is $sF(s) - f(0) = sF(s) - 2$.

As another illustration, if we have $d^2f(t)/dt^2$ with $f(0) = 0$ and $df(0)/dt = 0$ then the Laplace transform of $d^2f(t)/dt^2$ is $s^2F(s)$. If the function had $f(0) = 2$ and $df(0)/dt = 1$ then the Laplace transform of $d^2f(t)/dt^2$ is $s^2F(s) - 2s - 1$.

5 *Integrals*

The Laplace transform of the integral of a function $f(t)$ which has a Laplace transform $F(s)$ is given by:

$$\mathcal{L}\left\{\int_0^t f(t)\,dt\right\} = \frac{1}{s}F(s)$$

For example, the Laplace transform of the integral of the function e^{-t} between the limits 0 and t is:

$$\mathcal{L}\left\{\int_0^t e^{-t}\,dt\right\} = \frac{1}{s}\mathcal{L}\{e^{-t}\} = \frac{1}{s(s+1)}$$

The inverse transform

The inverse Laplace transformation is the conversion of a Laplace transform $F(s)$ into a function of time $f(t)$. This operation is written as:

$$\mathcal{L}^{-1}\{F(s)\} = f(t)$$

The linearity property of Laplace transforms means that if we have a transform as the sum of two separate terms then we can take the inverse of each separately and the sum of the two inverse transforms is the required inverse transform.

$$\mathcal{L}^{-1}\{aF(s) + bG(s)\} = a\mathcal{L}^{-1}F(s) + b\mathcal{L}^{-1}G(s)$$

To illustrate how rearrangement of a function can often put it into the standard form shown in Table 3.1, the inverse transform of $3/(2s + 1)$ can be obtained by rearranging it as:

$$\frac{3(1/2)}{s + (1/2)}$$

Table 3.1 contains the transform $1/(s + a)$ with the inverse of e^{-at}. Thus the inverse transformation is just this transform with $a = \frac{1}{2}$ and multiplied by the constant $(3/2)$ and so is $(3/2)\,e^{-t/2}$.

Expressions often have to be put into simpler standard forms of Table 3.1 by the use of partial fractions. For example, the inverse Laplace transform of $(3s - 1)/s(s - 1)$ is obtained by first simplifying it using partial fractions:

$$\frac{3s-1}{s(s-1)} = \frac{A}{s} + \frac{B}{s-1}$$

and so we must have $3s - 1 = A(s - 1) + Bs$. Equating coefficients of s gives $3 = A + B$ and equating numerical terms gives $-1 = -A$. Hence:

$$\frac{3s-1}{s(s-1)} = \frac{1}{s} + \frac{2}{s-1}$$

The inverse transform of $1/s$ is 1 and the inverse of $1/(s - 1)$ is e^t. Thus:

$$\mathcal{L}^{-1}\left\{\frac{3s-1}{s(s-1)}\right\} = 1 + 2\,e^t$$

Solving differential equations using Laplace transforms

Laplace transforms offer a method of solving differential equations. The procedure adopted is:

1 Replace each term in the differential equation by its Laplace transform, inserting the given initial conditions.

2 Algebraically rearrange the equation to give the transform of the solution.

3 Invert the resulting Laplace transform to obtain the answer as a function of time.

As an illustration, the first-order differential equation:

$$3\frac{dx}{dt} + 2x = 4$$

has the Laplace transform:

$$3[sX(s) - x(0)] + 2X(s) = \frac{4}{s}$$

Given that $x = 0$ at $t = 0$:

$$3sX(s) + 2X(s) = \frac{4}{s}$$

Hence:

$$X(s) = \frac{4}{s(3s+2)}$$

Simplifying by the use of partial fractions:

$$\frac{4}{s(3s+2)} = \frac{A}{s} + \frac{B}{3s+2}$$

Hence $A(3s + 2) + Bs = 4$ and so $A = 2$ and $B = -2/3$. Thus:

$$X(s) = \frac{2}{s} - \frac{2}{3(3s + 2)} = \frac{2}{s} - 2\frac{\frac{2}{3}}{\frac{2}{3}\left(s + \frac{2}{3}\right)}$$

and so:

$$x(t) = 2 - 2\,e^{-2t/3}$$

As a further illustration, the second-order differential equation:

$$\frac{d^2x}{dt^2} - 5\frac{dx}{dt} + 6x = 2\,e^{-t}$$

has the Laplace transform:

$$s^2X(s) - sx(0) - \frac{d}{dt}x(0) - 5[sX(s) - x(0)] + 6X(s) = \frac{2}{s + 1}$$

Given that $x = 0$ and $dx/dt = 1$ at $t = 0$:

$$s^2X(s) - 1 - 5sX(s) + 6X(s) = \frac{2}{s + 1}$$

Hence:

$$X(s) = \frac{\frac{2}{s + 1} + 1}{s^2 - 5s + 6} = \frac{2}{(s + 1)(s - 2)(s - 3)} + \frac{1}{(s - 2)(s - 3)}$$

We can simplify the above expression by the use of partial fractions:

$$\frac{2}{(s + 1)(s - 2)(s - 3)} = \frac{A}{s + 1} + \frac{B}{s - 2} + \frac{C}{s - 3}$$

Hence $A(s - 2)(s - 3) + B(s + 1)(s - 3) + C(s + 1)(s - 2) = 2$ and so $A = 1/6$, $B = -2/3$ and $C = \frac{1}{2}$. For the other fraction:

$$\frac{1}{(s - 2)(s - 3)} = \frac{D}{s - 2} + \frac{E}{s - 3}$$

Hence $D(s - 3) + E(s - 2) = 1$ and so $D = -1$ and $E = 1$. Thus:

$$X(s) = \frac{\frac{1}{6}}{s + 1} + \frac{-\frac{2}{3}}{s - 2} + \frac{\frac{1}{2}}{s - 3} + \frac{-1}{s - 2} + \frac{1}{s - 3}$$

$$= \frac{\frac{1}{6}}{s + 1} - \frac{\frac{5}{3}}{s - 2} + \frac{\frac{3}{2}}{s - 3}$$

The inverse transform is thus:

$$x(t) = \frac{1}{6}\,e^{-t} - \frac{5}{3}\,e^{2t} + \frac{3}{2}\,e^{3t}$$

Answers

Chapter 1

1   (a) Pressure on keys, display of calculated value, (b) electrical signal, sound, (c) radio signals, sound

2   Open – no feedback, closed – feedback, see Section 1.3.1 and 1.3.2

3   (a) Measurement – temperature sensor, controller – thermostat, correction – heater, process – water bath, (b) measurement – rotary speed sensor, controller – motor, process – shaft, (c) measurement – sensor of thickness, e.g. LVDT, controller – differential amplifier, correction – rollers, process – steel strip

4   See Fig. A.1

5   Measurement – thermistor with resistance-to-voltage converter, comparison – differential amplifier, correction - relay and heater, process – the enclosure being controlled; measurement – level probe, controller – relay and solenoid valve, correction – flow control valve, process – water tank, measurement – level probe

6   (a) $x - y$, (b) $x + y$, (c) $x - y - z$

7   Eg. (a) resistance temperature detector, (b) limit switch, (c) LVDT, (d) tachogenerator

8   (a) Push button activate, spring return: initially output 2 is at pressure, when button pressed output 2 is vented, (b) solenoid activated, spring return: initially output 2 is vented, when button pressed output 2 is pressurised,

9   (a) Pressing causes piston to move to left, release gives return, (b) pressing causes piston to move to left, release gives movement to right

10   See Figure A.2

11   (a) See Figure A.3(a), (b) see Figure A.3(b)

12   Ball operates a lever which opens or shuts a valve, depending on the height to which the ball is floating. The comparison element is the position of the lever, the control element is the lever, the correction element the valve, the process the water in the cistern and the measurement is the floating ball.

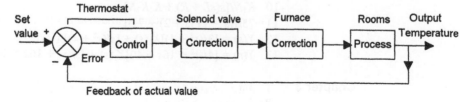

Figure A.1   *Chapter 1, problem 4*

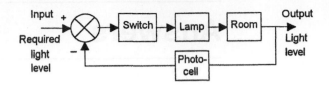

Figure A.2   *Chapter 1, problem 10*

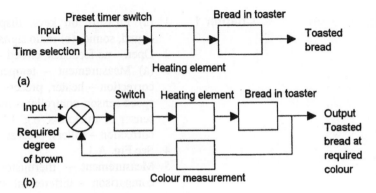

Figure A.3   *Chapter 1, problem 11*

Chapter 2

1 10 V

2 100

3 (a) $k_1 x = M\dfrac{\mathrm{d}^2 y}{\mathrm{d}t^2} + c\dfrac{\mathrm{d}y}{\mathrm{d}t} + (k_1 + k_2)y,$

(b) $T = I\dfrac{\mathrm{d}^2\theta}{\mathrm{d}t^2} + c\dfrac{\mathrm{d}\theta}{\mathrm{d}t} + k\theta,$

(c) $v = L\dfrac{\mathrm{d}i}{\mathrm{d}t} + Ri,$

(d) $p = c\dfrac{\mathrm{d}\theta}{\mathrm{d}t} + cpq\theta,$ where $\theta = \theta_\mathrm{o} - \theta_\mathrm{i}$

4 $5/[s(s + 1)]$

5 1.7

6 $2/(s + 1.2)$

7 $1/(1 + RCs)$

8 (a) $G_1 G_2 G_3$, (b) $G_1 G_2/(1 + G_1 G_2 G_3)$, (c) $G_1 G_2 G_3/(1 + G_1 G_2 G_3)$,
   (d) $G_1 G_2/(1 + G_1 G_2 H_1 + G_2 H_2)$

9 (a) $1/[s(s + 2) + 2]$, (b) $K/[s(s - 1) + K(s + 2) + Ks]$,
   (c) $2/(s^2 + 9s + 6)$

10 $K_1 N/[s(sL + R) + K_1 K_2 N]$

11 $75/[(s + 1)^2(s + 2) + 7.5]$

12 $Y(s) = [2/(s^2 + 4)]X_1(s) + [s/(s^2 + 4)]X_2(s)$

13 $Y(s) = [4/(s^2 + 16s + 40)]X_1(s) + [s/(s^2 + 16s + 40)]X_2(s)$

Chapter 3

1 $3/s$

2 2

3 $5/s^2$

4 e^{-5t} V

5 (a) $(5/3)(1 - e^{-3t})$ V, (b) $5e^{-3t}$ V
6 (a) $6(1 - e^{-t})$ V, (b) $6e^{-t}$ V
7 (a) $2(1 - e^{-2t})$ V, (b) $\frac{1}{2}[t - \frac{1}{2}(1 - e^{-2t})]$ V
8 (a) $5/(s - 1) - 4/(s - 2)$, (b) $4/(s + 1) - 3/(s + 2)$,
   (c) $2/(s + 1) - 3/(s + 1)^2$
9 $24 - 12e^{-2t} - 4e^{-4t}$
10 $8e^{-2t} - 8t\,e^{-2t}$
11 (a) Critical, (b) overdamped, (c) underdamped
12 $-1.5 + 3.0t + 1.5e^{-2t}$
13 $0.5 - e^{-t} + 0.5e^{-2t}$
14 $0.5(e^{-t} - e^{-3t})$
15 $0.5(1 - e^{-10t})$
16 10, 0.05
17 Underdamped
18 Critically damped
19 1/53 s

Chapter 4
1 (a) 9.2 s, 8.8 s, (b) 2.3 s, 2.2 s, (c) 0.77 s, 0.73 s
2 69.1 s, 65.9 s
3 8.7 s, 20 s, 19.1 s
4 (a) 10 rad/s, 0.2, 0.16 s, 53%, 2 s, (b) 7 rad/s, 0.29, 0.23 s, 39%,
   2.0 s
5 4 rad/s, 0.63, 0.51 s, 7.8%, 1.6 s
6 2.6
7 Stable (a) and (c); oscillatory (b) and (d)
8 (a) Stable, (b) unstable, (c) unstable
9 (a) Stable, (b) unstable, (c) stable
10 (a) Stable, (b) unstable

Chapter 5
1 (a) 2, 90°, (b) 2, 180°, (c) 2.2, 26.6°
2 (a) $3/(5 + j\omega)$, (b) $7/(2 + j\omega)$, (c) $1/(10 + j\omega)(2 + j\omega)$
3 (a) 1.34, −26.6°, (b) 1.06, −45°
4 See Figure A.4
5 See Figure A.5

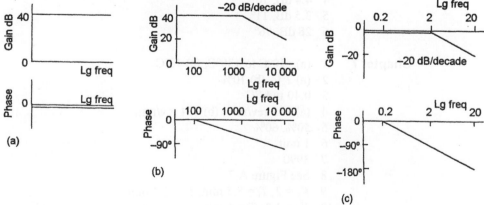

Figure A.4   *Chapter 5, problem 4*

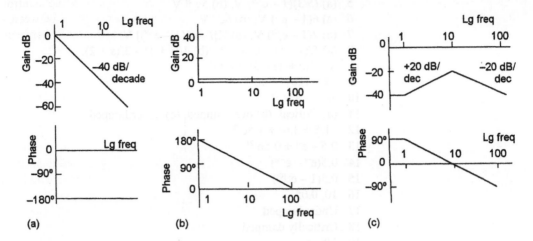

(a)   (b)   (c)

Figure A.5   *Chapter 5, problem 5*

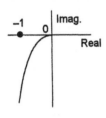

Figure A.6   *Chapter 6, problem 1*

6 (a) $1/s$, (b) $3.2/(1 + s)$, (c) $2/(s^2 + 2\zeta s + 1)$,
   (d) $3.2 \times 100/(s^2 + 2\zeta s + 100)$, (e) $10/(s^2 + 0.8s + 4)$
7 $16 \times 15/(s + 15)$
8 1.82, 50°
9 (a) Stable, (b) about 36 dB, (c) about 80°
10 (a) Stable, (b) about 15 dB, (c) about 43°
11 (a) Stable, (b) about 15 dB, (c) about 32°
12 About 5.7 dB, reduced by factor of 1.9
13 About 34°, 1.7 Db, reduced by factor of 1.2
14 55°
15 $(0.38s + 1)/(0.13s + 1)$
16 $(10s + 10)/(80s + 1)$

Chapter 6   1 See Figure A.6
2 $(-1, j0)$ point not to be enclosed
3 8.0 dB, 20°
4 4.4 dB, 10°
5 3.5 dB, 11°
6 28 dB, 76°

Chapter 7   1 (a) −10°C to +30°C, (b) 40°C
2 (a) 2°C, (b) 10%
3 0.40 mA/°C
4 (a) 255 rev/min, (b) 324 rev/min
5 50%, 60%
6 1 unit
7 3990
8 See Figure A.7
9 $K_p = 2$, $T_i = 5.5$ min, $T_d = 1.4$ min
10 $K_p = 1.3$, $T_i = 6$ min, $T_d = 1.5$ min
11 $K_p = 0.62$, $T_i = 3$ min, $T_d = 0.75$ min

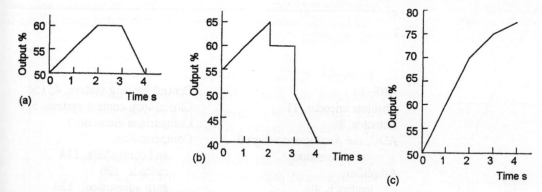

Figure A.7 *Chapter 7, problem 8*

12 $\dfrac{10(K_{d}s^{2} + K_{p}s + K_{i})}{s(s = 5)(s + 10) + 10(K_{d}s^{2} + K_{p}s + K_{i})}$

13 (a) –40°, (b) 15°

14 (a) $4 - 4\,e^{-0.5t}$, (b) $5[1 - 0.997\,e^{-0.05t}\sin(0.63t - 1.49)]$

Index

Printed and bound by CPI Group (UK) Ltd, Croydon, CR0 4YY

03/10/2024

01040336-0013